The R.A.M.S. Library of Alchemy

Volume 17

Philosophical Furnaces

by
Johann Rudolph Glauber

R.A.M.S. Publishing Company

Philosophical Furnaces

By

Johann Rudolph Glauber

Produced by

Restorers of Alchemical Manuscripts Society
1983

R.A.M.S. Publishing Company

R.A.M.S. Publishing Company
1112 Alston Court
Waynesboro VA 22980

Philosophical Furnaces
Copyright © 2017 Althea Productions LLC
All Rights Reserved

http://ramsalchemy.jimdo.com

All rights reserved. No part of this publication may be reproduced or transmitted in any form or by any means, electronic or mechanical, including but not limited to any information storage and retrieval system, without written permission from Althea Productions LLC.
Reviewers may quote brief passages.

First Edition January 2017

ISBN-13 **978-1542383011**
ISBN-10 **1542383013**

Image Processing by Philip N. Wheeler

This book is sold for informational purposes only. Neither the publisher nor the editor shall be held accountable for the use or misuse of the information in this book.

Printed in the United States of America

Table of Contents

Disclaimer 7

Introduction 9

To Edmond Dickenson, M. D. 15

The Preface to the Reader. 19

Explication of the Figures of the First Furnace ... 35

Philosophical Furnaces. 39

First Part of Philosophical Furnaces 45

Second Part of Philosophical Furnaces 103

Third Part of Philosophical Furnaces 259

Fourth Part of Philosophical Furnaces 311

Fifth Part of Philosophical Furnaces 381

An Appendix 417

A Word from the Publisher 423

The R.A.M.S. Library of Alchemy 424

Dedicated to Hans W. Nintzel,
American Alchemist
and
Founder of the
Restorers of Alchemical Manuscripts Society
(R.A.M.S.)

This book is published with permission from
The Estate of Hans W. Nintzel.

Disclaimer

Liability: The publisher does not warrant or assume any legal liability or responsibility for the accuracy, completeness, or usefulness of any information, apparatus, product, or process disclosed. The publisher makes no representation as to the accuracy or completeness of the contents of this book and specifically disclaims any implied warranty of merchantability or fitness for a particular purpose. No warranty may be created or extended by written sales materials or sales representatives. You should obtain professional consultation where appropriate. The publisher shall not be liable for any loss of profit or other commercial or personal damages, including but not limited to special, incidental, consequential, or other damages.

Introduction

Philip N. Wheeler

In *Philosophical Furnaces*, Rudolph Glauber presents detailed plans for four furnaces, each designed for a different purpose. The primary process described is distillation. The text contains instructions for producing various medicines, and other products such as artificial gems, magnifying glasses, strong porcelain utensils, fertilizer, wine, vinegar, Potable Gold, and much more. Although a few key details may have been left out, they are said to be presented in Glauber's other works. This book covers an incredibly wide range of early chemical experiments. Included are 6 pages of excellent engravings restored from the original printed work produced by Christopher Packe in 1689.

"The Complete Works of Rudolph Glauber" was the largest R.A.M.S. project ever undertaken by Hans W. Nintzel. Comprised of over 2,000 pages, it was a massive effort. Many members of R.A.M.S. contributed their labor to this project, which was finally completed in 1983. This book, *Philosophical Furnaces*, is but a small part of that effort. The original production by Hans has been divided to achieve reasonably sized bound volumes.

The overall title of both parts of the original R.A.M.S. release was, *Secrets of Medicine and Alchemy*. This Volume includes all five parts of *Philosophical Furnaces*. The illustrations are from the original 1689 printed edition.

My goals in editing were:
- Preserve the meaning of the original translated text,
- Improve overall readability, and
- Correct as many typographical errors as possible, while preserving all words whose meaning is not immediately without question.

A. Ist der Ofen.
B. Das distillier gefäs in dem Ofen.
C. Das loch mit einem falz dardurch das swänge eingetragen wird. D. Die Zange mit deckel darmit das gefäs nach dem eintragen geschlosse wird. E. Ein Eisen löffel da mit der schwan eingetragen wirdt F. Die recipienten. G. Die banck darauf die recipienten hgen. H. der Laborant.

A. Ist der Ofen wie er in der arbeit stehet. C. Ist das obertheil des Ofens   G. der Deckel dar mit der Ofen oben gedecket wird
B. Ist der Unterste theil des Ofens wie er D. Der rost in dem Ofen. E. der Herd.   wann kohlen darin sein vnd in voller arbeit
offen ohne die vorder wand anzusehen. F. der Muffel so auff den herd gehöret   stehet. H. das rauchfänglein an dem Ofen.

The Works of the
Highly Experienced and Famous Chymist,
John Rudolph Glauber;
containing,
Great Variety of Choice Secrets
in
Medicine and Alchymy
In the Working of Metallick Mines, and the
Separation of Metals,

Also

Various Cheap and Easy Ways of making SaltPetre, and
improving of Batten-Land, and the
Fruits of the Earth.

Together with many other things very profitable for all the Lovers of Art and Industry. Translated into English, and Published for Publick Good by the Labour, Care and Charge, of Christopher Packe, Philo-chymico-Medicus.

London, Printed by Thomas Milbourn, for the Author, and are to be sold at his House next Door to the Gun in Little-Moorfields; by D. Newman at the Kings-Arms in the Poultry, and W. Cooper at the Pellicam in Little Britain. MDCLXXXIX.

To Edmond Dickenson, M. D.

To The

Honoured, and Truly Learned,

Edmond Dickenson, M. D.[1]

Physician to the King's Person and Family.

The Art of Chymistry (Honoured Sir), although in its Speculations most Noble and Delectable to a Philosophick Mind, and in its Practice highly Inservient, and Beneficial to Mankind; yet has it not escaped the Obloquies, and false Imputations of Detractors, and Calumniators, who either through Ignorance, Idleness, or Envy (or all of them conjoined) have made a false Representation of this most Noble Art to the World, and endeavoured to set Mankind at the greatest distance from that which is its highest interest to court. For which cause such Writings as Promulge, and offer at the advancing of the CHYMICAL ART, stand in need of such a Patron as is able to defend them against all the Cavils of Pride, Envy, and Ignorance.

And if the Exquisite Parts, and Profound Learning in the more Abstruse Philosophy, together with a Long, and Indefatigable Scrutiny and Labour in the Chymical Art, accompanied with a happy Practice in the Honourable Faculty of Physick, be fit Accomplishments to Entitle one a MECENAS of this Art; then are those Excellencies all met and

---

[1] Edmond Dickenson (1624-1707) was an English royal physician and alchemist. -PNW

Concentered in yourself, as is evident to the whole World by your Curious and Learned Epistle to MUNDANUS, and his Answer to it, which answer will be a Lasting Testimony of your great Worth and Merit.

For certainly, Sir, it is no small evidence of your Worth and Abilities in the PYROTECHNIC ART, that a Philosopher who had been more than forty years an ADEPT, in all that time should not find three Persons, besides yourself, whom he thought worthy to make certain of the truth of what they sought, and aspired after; and yet gave you an Ocular Satisfaction and Certitude of that which thousands have desired to see, but could not: And further seriously professing, that if he had had the same liberty from his Master, that some ADEPTS enjoy, that he would have revealed to you the whole Secret.

These things have induced me humbly to offer this Book to your Patronage, not doubting but under your Name and Protection, it will be able to overcome many Difficulties, and obtain a free passage in this our English World, to the benefit and advantage of many well-disposed persons, who seek after Honest, Profitable, and Commendable Arts, which I am fully persuaded was the chief end of the Author in Writing: and I am sure is mine in Translating his Works. You are thoroughly acquainted with Glauber's Writings, you know his Menstruums, and his Medicines, and are able to attest the truth of what others may account false and impossible. As for such of them as concern the higher Classes of Chymistry, I shall say nothing (being yet but *ad Corinthum vergens*) but commit them to your Mature Judgement, and protection, humbly craving your pardon for this my presumption, and for what Errors or Oversights I may have committed in this Work; and desiring your Favorable Acceptance of these my poor Endeavors. I take leave to conclude with a passage of the abovementioned Excellent MUNDANUS. I am fully persuaded, that by the Blessing of God upon your Sagacious Labors, you will at length obtain that which will abundantly Compensate your Pains and Cost. To which I adjoin my own hearty Wishes; and that after you have been as happy in this World, as true Philosophy can make a Man, you may be Eternally Happy in that which is to come. I am

    Sir,
    An Honorer of
    Your Name and Learning,
    Christopher Packe.

The Preface to the Reader.

That the Art of Chymistry is very useful and highly serviceable in Physick, Chirurgery[2], Husbandry, and Metallick Arts, is long since evinced by the Excellent Mr. Boyl[3] (the Honor both of our Age and Country) in his Experiment at Philosophy, or **Philosophick Essays;** who in Essay I. and II. shows that the Examination of the Juices of Human Bodies, by the Art of Chymistry, may illustrate their Use and Nature. And that by it may be Explicated the Nature of our several Digestions, and their Attractions. And afterwards Cap. VIII. page 191. speaking of the advantages that Chymistry affords to the Therapeutic or Curative part of Physick, (which is the chief and principle) and to which all the other parts are subservient) is pleased to express himself thus: I cannot but think that if Chymistry did no more than assist us, by the resolution of Bodies, to extricate their more active parts, and partly by such Resolutions, and partly by associating Bodies together, to alter the former Texture of Natures productions, or present us with new Concretes of new Textures: by this very means, if Men want not Curiosity and Industry, to vary and prosecute Experiments, there must necessarily arise such a store of new and active Medicines, that in all probability, many of them will be found endowed with such virtue as have not been (at least in that degree) met with, in the usual Medicines, whether

---

[2] Surgery. -PNW
[3] Robert Boyle (1627-1691), widely regarded as the first modern chemist. -PNW

Simple or Compound, to be bought in Apothecary's Shops; and consequently, even without any notable discovery, or improvement of Principles, Chymists (even as Matters now stand with them) may considerably add to the Pharmaceutical part of Physick. But if the Operations of Chymistry were seriously enquired into, and thoroughly understood, I make little doubt, but by a skillful Application of them, and especially by a series of them, in a Rational and Orderly way succeeding one another, there may be found out a great many preparations of Remedies, both very different from the common Ones, and far more Noble then they. And presently after he adds: That if we had but a few Potent Menstruums to dissolve and unlock Bodies with, I scarce know what might not be done in Chymistry. Then further in that Essay where he treats of the usefulness of Chymistry to the Empire of Man over the Inferior Works of Nature; he proceeds to show that Chymistry is very serviceable to Husbandry in all its parts, and to other professions that serve to provide Men with Food or Raiment, or do otherwise minister to the Necessities or Accommodations of Life, as Bakers, Brewers, Dyers, & etc.

    Thus far this Learned Philosopher: To which I shall only add this, That if when he wrote those Essays, Chymists were able to contribute so much to the Necessities and Conveniences of Mankind, when Chymistry was but young in England, and but few Chymists who were accurate in their Operations, and perhaps fewer who had any competency of Learning, or so much as lightly Tincted with the Hermetic Philosophy; if, I say, that it discovered so great a light when it had but newly ascended our Horizon, and was, as I may say, but in its infancy, what assistance may now be had from it, when (notwithstanding all the obstacles, and unkind usage it has met withal) it is grown to a more virile Age and Vigor: But although Chymistry be much enlarged, and advanced in England, in respect of the Numbers, and Qualifications of the Lovers and professors of it; yet are not Chymists free from pressing Disadvantages, not having the freedom of administering their own Medicines, how powerful and salutiferous so ever, and otherwise adapted to the necessities of the Sick, than the common Apparatus of Physick. So, that as the Case now stands, the help and Succor which the Sick and Diseased receive from Chymical Physick, is but very small to what they might have, if knowing Chymists had the freedom of exercising that Art in all its parts, which with much Industry, Labour, and Costs, they have been solicitous to attain. But when this discouragement of ingenuity and Obstacle of the public good, shall become more apparent to those in whose power it is to redress it, I do not doubt but it will meet with a Remedy.

But now, to give some account of my present undertaking. I have at length (by God's help, and the assistance of my Subscribers) finished my Translation of Glauber's Works, and here present it to the Reader, in the English-Tongue. How well I have performed it, I must submit to the judgement of others: I could have been very glad to have seen it done by some abler hand; but when I have heretofore proposed the doing but of some parts of it to those whom I knew might easily have accommodated English Artists therein; telling them that I wondered so Excellent an Author, should be so long extant, and that none should unveil him of his Latin and German Coverings, and put him into an English Dress.

I have had for an answer, that this Age was not worthy of it; so, that it seems to me, that the Providence of God had reserved it for fitter times, although to be done by one of the meanest of the Sons of Pyrotechny. But this I can say, that I have acquitted myself in this matter, as well as the slenderness of my Parts, weakness of Body, and the necessary Affairs of my Laboratory would permit me; but;

Ubi desint Vires, acceptanda est Voluntas[4]. I desire the Lovers of Chymistry to accept my Labors, with the same good will that I have undergone them, having no other end but to serve my country. And I hereby return thanks to all those generous spirited Gentlemen and others, who have Subscribed to, and promoted this Work, without whose assistance (the Charge being very great, as well as the labor to me, almost insupportable) it must yet have remained hid and unserviceable to the English Reader. But I am in an especial manner obliged to that spirited Gentleman (whom I ought to name, were it lawful to do it without his leave) who freely offered me and put into my hands a not inconsiderable part of the Materials for this Work, which part, also had been more considerable than it was, had not the Spirit of some, (who unjustly hindered it) been as Mean and Sordid, as his was Generous. But that loss was, in part, made up to me, by a well-minded Artist, to whom I also return Thanks.

I have Printed this Book upon far better and larger Paper than I proposed to do it in; for as the time of setting forth my first Proposals, I had not the German Pieces, but when they came in my hands, upon a more accurate Computation of the matter, I

---

[4] Where there is no strength, to accept is will. -PNW

found that if I should go on to do the Work upon the Paper I had proposed, the Book would swell to too great a thickness for its breadth and length, and not be only ill shaped, but inconvenient to be read. By this means my Subscribers have a much better Book than I promised them, although the Charge has also been Considerably greater to me, than I at first expected.

    The Reader has all here in one Volume which Glauber ever Printed[5], as far as I can find upon diligent Enquiry at Amsterdam, where all his Writings were Printed, and where I purchased the Original Copper Plates belonging to them. But whereas, as it is said in the Explication of **Miraculum Mundi.** That the Cut there described was not Printed in the Latin Copies, nor to be found among the Original Plates; yet notwithstanding, I was not willing that the Work should go without the Figure of so useful a Furnace as that is, for the Torrefying, or Calcining of Ores, and separating, and depurating their Metals, for which reason I have caused it to be Delineated and Printed with others before the continuation of **Miraculum Mundi.** I have also procured from the hand of another friend, who is a Lover of Art, the Draught of the Refrigeratory, Furnace, or Instrument, which serves for the making the Mercury of Wine, purifying, and fixing of Argent-vive, Antimony, Sulphur, & etc. and many Other uses which an Ingenious Artist will find out. This Furnace the Author always endeavoured to conceal, but describes it in some part in the beginning of the sixth part of the **Spagyrical Dispensatory,** to which Description I have added the

---

[5] Although the original printed English edition was one huge volume, it was necessary to split it into several volumes for the R.A.M.S. Library of Alchemy. -PNW

Figure. The Figures of the several Vessels and Instruments belonging to the Fifth Part of the **Furnaces**, are referred to at the beginning of the Fourth Part, but since, for the better orders sake I have placed them before the said Fifth Part.

These Twelve following Treatises were never Printed in Latin, but in the German Tongue only, viz. The **Third, Fourth**, and **Fifth Centuries**; the Second and Third Appendixes to the Seventh Part of the **Spagyrical Dispensatory**. The **Book of Fires. Proserpine. Elias the Artist.** The **Three Firestones. The Purgatory of Philosophers. De Lapide Animali. The Secret Fire of Philosophers**, all which I have caused to be Translated (myself being ignorant of the German Tongue) by a person well skilled both in High-Dutch, and also in Chymistry, whereby I hope this Book will not be altogether unserviceable even to the Learned; besides, all the Works of this Author that are very difficultly (if at all) to be met with at any Book-sellers Shop in London, and those that are, at a dear rate: For when I had entered upon this Translation, I was forced to send to Amsterdam to have all the Latin pieces complete.

The Author in many places refers to his **Opus Saturi, Opus Vegetabile**, and the **Concentration of Heaven and Earth**, which Treatises, I am assured, were never printed (at least under those Titles) which also seems to be manifest from his Epistle to the **First Century**, or General Appendix, wherein he inculcates, that for want of time, he had inserted the sum of them all in that Treatise. He also mentions a Seventh part of the **Prosperity of Germany**, in the Preface to the Second Part of **Pharmacopoeia Spagyrica**, which was never Printed under that Title, but I am induced to believe it is in the **Novum Lumen Chymicum**, as partly appears by comparing it with the foresaid Preface. And it is evident that in some parts of his Writings he has mentioned a Treatise by one Name and afterwards Printed it by another, as, **The Testimonium Veritatis**, which was afterwards Printed by the Name of **Explicatio Miraculi Mundi**. As for the **Opus Saturni**, I have heard that there are some Manuscript Copies of it, and had hopes of obtaining it from two of several hands, but both failed me. I have been also informed, that there are **Five Centuries** in Manuscripts more than I have Printed, but could never understand in what hands they were, except one of them, viz. the sixth, the proprietor of which would not be so kind to let me have it to print.

I have (by the advice of an Honourable Person) left out the Author's Religious and Moral Digressions, where I could do it without prejudice to the matter; as also his Apologetical Writings, except his **Apology against Farmer**, which I have printed, for as much as it is intermixed with many profitable Secrets, which perhaps, he would not have published, at least not at that time if they had been, as it were, extorted from him by the ill Treatment of that Ungrateful Man.

I could not place the several Treatises in that order which the Author published them, without breaking the order of the several parts, as of the **Miraculum Mundi**, **Spagyrical Pharmacopoeia**, and **Prosperity of Germany**; for being many years in publishing, they were done promiscuously, but how they succeeded one another so far as the Nature of Salts, the Reader may satisfy himself in the Preface to that Treatise. And as his Writings were published by piece-meal, so are the principal Secrets he teaches, scattered up and down in divers parts of them, in one place he treats of a thing obscurely, or but in part, in another place of the same thing openly in that part which he had veiled in the other. Sometimes he declares a Process very openly, omitting only some small Circumstances, or Manual Operation, which would seem to many either impertinent, or not necessary to be done, when notwithstanding, the business will not succeed without it. An instance of this may be given in his **Sal Mirabilia**, whose preparation he teaches obscurely in the **Nature of Salts**, but more openly in the Second Part of **Miraculum Mundi**. In the **Nature of Salts**, and in the Sixth Part of the **Pharmacopoeia Spagyrica**, he teaches how to dissolve Gold therewith, and thence to make a kind of Aurum Potabile, but wholly omits the adding of a certain Vegetable Sulphur, without which, the work will not answer the Description; this Defect he supplies in the **Second Century**, after a twofold manner, the one not obvious to every man's apprehension, I mean the intent of the Author, viz. in those Processes where he shows the making of a Vegetable Sulphur; but the other shows the necessary Manual Operation in plain and open words. And this he has done with all his Secrets on set purpose, that they should be found out by none but the Industrious.

And this has given occasion to many, who have not taken pains to read him with diligence, or not being experienced in Operating, to reproach him for an obscure, yea, even for a false Writer, because they have made two or three Superficial, or Unskillful Trials of his Processes, which have not succeeded according to their Expectations, when indeed, the faults were in themselves, either in not perceiving the Authors intention, or their own want of skill in rightly managing the Operation: And I know some persons that sometime scarce said Glauber had been too dark in his Writings, who now think he has written too plainly.

But having mentioned this, I will here (for the sake of those Country Gentlemen, who have subscribed to this Work) a little Elucidate the Authors Process about the inversion of Common Salt, with Lime, for the enriching of Poor and Barren Land. He indeed speaks of several Saline Preparations, which greatly promote the fertility of the Earth, but this with Common Salt, and Lime, is the cheapest of all, and is most easy to be done, for any Plow-man, or Laborer, having but once seen it done, may be presently able to manage it. The sum of it is, that Common Salt be turned from its sharpness, into an Alcalizate Nature (which is hot and fat) which then by its Magnetic force will attract from the Air a Vivifying, Fructifying, Salt-nitrous power, and long retain it in the Earth, which is the cause of all Growth and Vegetation, as the Author shows in the Continuation of **Miraculum Mundi**, and many other places; but gives the Process of the preparation in plain and open words in the Appendix to the Fifth Part of the **Prosperity of Germany**.

Neither is the practice of preparing either the Land or the Seed, in order to better the Crop, altogether Novel, as may be partly seen in Virgil, Georgic Lib. 1. where he says,

Semina vidi equidem multos medicare ferentes,
Et Nitro prius, & nigra perfundere amurca:
Grandior ut foetus filiquis fallacibus esset,
etc.

Which in English may sound thus:

> Some have I seen their Seeds to sow prepare,
> With Nitre and Oil-Lees, for they by care
> Will grow far greater, and be sooner ripe,
> etc.

The Lime must be spread upon the ground, where no Rain can come to it, till it slake itself by the Air, and fall into a Powder; of this Powder you are to take four hundred weight to one hundred weight of any common foul Salt, which is too impure for the use of the Kitchen, where such may be had, otherwise clean Salt, (for that will be cheaper than Dung) the Salt and Lime are to be well mixed, and then moistened with such a quantity of Water, (or rather Urine where it may be had) as will bring the Lime and Salt mixed, to the Consistency of a stiff Mortar. Of this Mass, Balls are to be made about the bigness of ones fist, and laid under a Shed, or Hovel to dry; being dried, they are to be burnt in a Kiln as Lime is, so that the Balls may be red hot for an hour at least; or where no Lime—Kiln is near, they may be burnt by building a Pile in the Field, first with a Lay of Wood, then a Lay of Balls, then Wood again, and so till the Balls are placed fit for burning. When the Balls are burnt, they are to be again placed upon a Floor under a Shed, or Hovel, where they may be exposed to the Air, but kept free from the Rain, and if you break them with a Clod-beater presently, the Air will the sooner act upon them, and cause them again to fall into a Powder; which Powder may then be carried out and spread, or rather sowed out of a Seeder, thicker or thinner as the Land shall require. Provided this be done in the beginning of Summer about the time of Fallow, for that being many Months before the Seed is to be sowed, the fieriness of the rich Compost will be

Contempered by the Air and the Earth, and changed into a Nitrous fatness, which joining itself with the Earth, is again Magnetically attracted by the Seed when it is sown, whose growth is thereby swiftly promoted, and its Multiplication much augmented. But if any should cast this Matter upon his Land soon after it is burnt, and presently after that should sow his Seed, instead of having a greater Crop then he used to have, he would have a less, or perhaps none, that Year, but the next Year, and so on for many Years, the same Land would bring forth Plentifully. Therefor it is necessary, that this Matter should lie six or seven Months spread upon a Floor, and now and then turned with a Shovel, as you turn Malt, that it may be Contempered, and Animated by the Air; or be cast upon the Land so long before the Seed be sown. The reason is the very same with Dung, for none takes fresh Dung and spreads it upon his land when he is about to sow his Seed, for if he should, his Seed would burn up; but the Husbandman lets his Dung lie some time to rot, as he calls it, after which he lays it on his Land, and lets it lie spread some time before he plows it in, and all this is but to Contemper the heat of the Animal Salt contained in the Dung, and turn it into a Nitrous Nature. This much I thought good to say about this Matter in the plainest words, least any, not thoroughly understanding the Authors Intention, should err in the first Experiment, and so unjustly blame the Author, and forbear themselves and deter others from prosecuting that easy Practice, which I am confident, if rightly managed, will bring much profit to many persons in this Nation. This must also of necessity be a profitable Work to those who will undertake it upon the account of making of Salt-petre; especially to such as understand the

Nature and Generation of that Excellent salt, which is of such incomparable use in the Preparation of Medicines, separating of Metals, and in many Mechanick Arts.

    Now for as much as in this Work Sal Mirabilis, Spirit of Nitre, and Spirit of Salt, are recommended to very many uses, and everyone that has a mind to make Experiments with them may not have the knowledge, or the convenience of preparing them, I hereby signify, that I intend (God willing) to prepare and keep by me the Author's Sal Mirabilis of both sorts, that peculiar Spirit of Salt, which he commends against the Scurvy and other Diseases, and also to keep Beer from souring in the Summer, in the **Consolation of Navigators**. His Panacea of Antimony, and Golden Panacea, spoken of in the Second Part of the **Pharmacopoeia Spagy.**, the Explication of **Miraculum Mundi**, and divers other places. His **Aurum Diaphoreticum**, also the Tincture of Gold, or **Aurum Potable**, are described to be made of the Irreducible Blood of the Lion, in the Sixth part of the **Spagyrical Pharmacopoeia**, Chap. 22. These I propose constantly to keep by me for the accommodating of Physicians, and others, who shall have occasion to buy them. Those are Excellent Medicines, and such as a Physician may have some confidence in; and indeed, this Book contains a great variety of such Medicines as will get a Physician Honor, which (I hope) will be tried by all those who delight to do good, and be brought into use for the general Help and Comfort of the sick. For I freely confess, that if I have anything in Medicine, beyond what is commonly known, I have had the Foundations of it from this Author; and if God shall please to grant me life to a fit time, I doubt not but I shall from those Foundations be able to raise such a Superstructure as shall testify the truth of his Writings, and powerfully evince the Worth and Excellency of Chymical Medicines, and that demonstratively in matter of Fact, viz. by the Curing of both Acute and Chronic Diseases.

And now by way of Conclusion, I have only one thing more to add; and that is a Request to all the Ingenious Lovers of Chymistry, that they would not occasion this Work, which I have undergone with so much Labour, and loss of time from my private Concerns, merely for the good of others, to redound to my own hurt; my meaning is, that I might not be put to the charge and trouble of Letters about Curious Enquiries, wherein I am to have not the least profit: This I mention, because I have had divers such Letters come to my hands since I have been about it, and that sometimes two or three being very long ones with many Queries, in one Week. Now should this continue, and I endeavor to satisfy all the Doubts, and gratify all the Curiosities of all such non-considerate persons, truly I should have no time besides what this would take up, to provide for myself and Family. But notwithstanding what I have said, if any Ingenious Person shall stand in need of my Assistance, in preparing of anything for him, or otherwise, wherein I may have a reasonable recompense for my Time and Trouble, I will be ready to give him the best assistance I can. For I am now but just ready to receive a Writ of Ease from three Years of daily Labour and care about this Work, and I would be willing to enjoy it sometime, that I might again with diligence apply
myself to my Laboratory, the effects of which, if God shall see good, may at one time, or other, show themselves to the World. In the meantime, I wish all Honest and Ingenious Lovers of the Spagyrick Art,

good success in their Studies and Labors, that thence the Penuries and Miseries of Mankind, especially of the sick, may be effectually remedied; that they may Cooperate as Instruments with the great ends and providences of the Almighty, to bring about that time, in which God shall be Glorified all the World over, and Men live in a more serene and tranquil condition than yet they have done, which shall always be the Desire and Prayers of him that is a Lover of Pyrotechny, and Honorer of all true Artists.

From my House, next Door to the Sign of the Gun in Little Moor—Fields, 1688.

             Christopher Packe.

The
Explication of the Figures of the First Furnace
In all its parts.

Fig. I. E. The first subliming-Pot, which is set into the upper hole of the Furnace. D. The upper hole of the Furnace. F. The Second Pot. G. The third. H. The fourth.

Fig. II. A. The ash-hole, with the wideness of the Furnace. B. The middle hole, by which the Coals and the Matter to be distilled, are cast in. C. A Stopper of Stone, which is to stop the said hole after casting in the matter. D. The upper hole with a certain false bottom, which is to be filled with Sand. E. The Cover of the upper hole, which is put on after the putting in the Coals and Materials. F. A Pipe going out of the Receiver, and joyned to the first Pot. G. The first Receiver. H. The second. I. The third. K. A Stool on which the first Receiver stands, having a hole in the middle, through which the Neck of the first Pot, to which a dish is annexed, passes. L. The Dish through the Pipe whereof the refrigerated Spirits distill. M. A Receiver into which the Spirits collected in the Dish do flow. N. A Screw to be raised higher at pleasure for better joining the Receiver to the Pipe, and it goes through the Stool. O. The place of the Pipe for Distilling the Spirit of Vitriol and Allome. P. A Grate consisting of two strong cross Iron Bars, fastened in the Furnace, and four of five more less, that are moveable, for the better cleaning of the Furnace.

Figure III. G. The first crooked Pipe fitted to the Pipe of the Furnace. F. The Pipe of the Furnace. H. A Receiver fitted to that Pipe, and set in a Tub of water, for accelerating the Operations: which Receiver has a Cover with two holes, through the first whereof goes a single crooked Pipe, and through the other two crooked Pipes, whereof one goes into the Receiver, as did the single, and the other out of the Receiver. H, into H. H. I. The Tub of Water. M. A third Pipe. By this way Flowers are sublimed, and Spirits distilled speedily, and in great quantity.

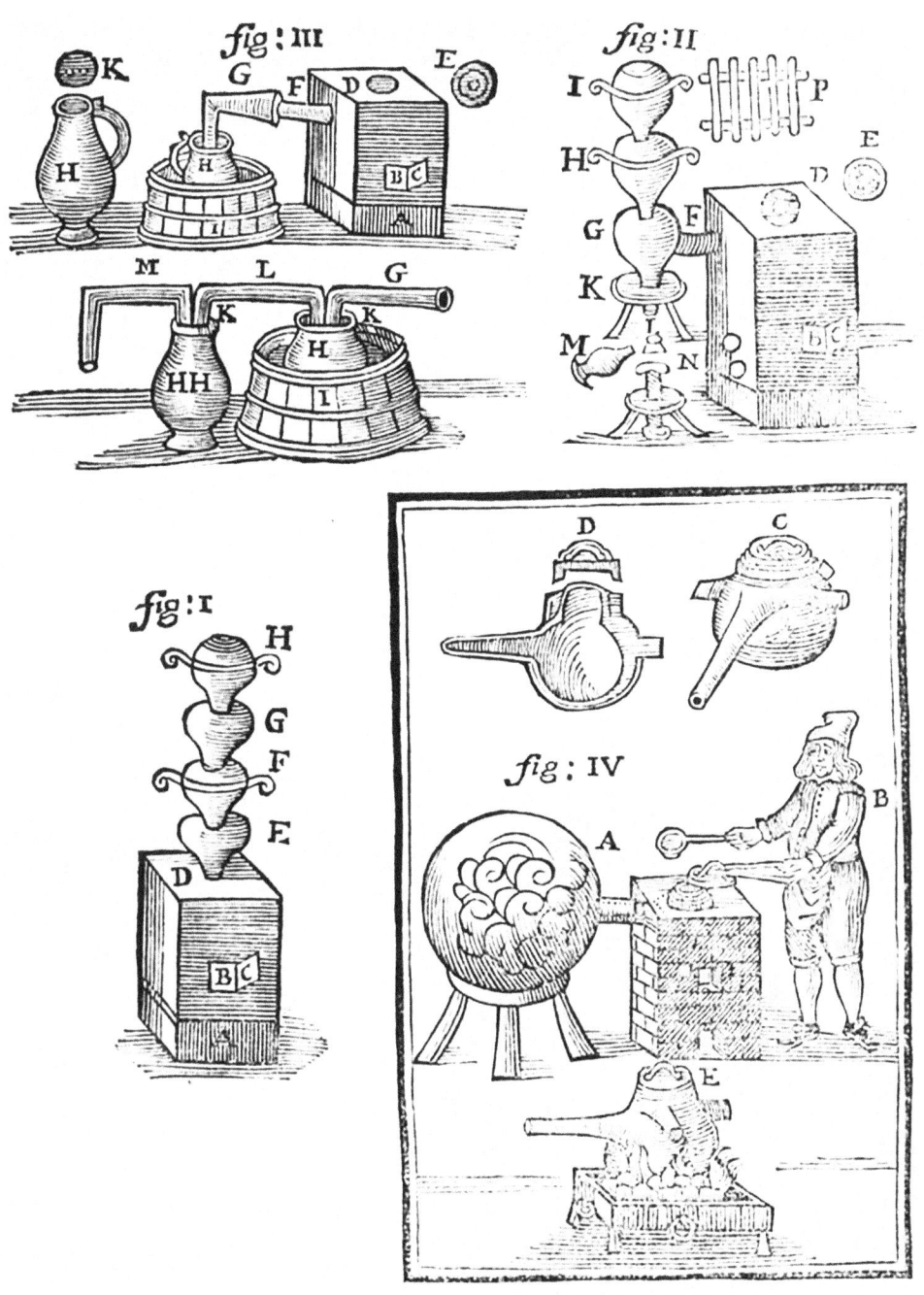

The Second Furnace.

The

The FIRST PART

OF

Philosophical Furnaces.

Containing a new Art of making SPIRITS, OILS, FLOWERS, and other Medicaments, by the help of the first of those Furnaces, after a very easy and peculiar manner out of Vegetables, Animals and Minerals; With their Chymical and Medicinal use.

A Preface to the Courteous Reader.

I have hitherto reserved to myself as Secrets, some peculiar Furnaces and compendious Ways of Distilling, which with diligent study and speculation I found out some few years since, by which many excellent Works, impossible to be done by the vulgar Art, may be performed; but now at last I have considered with myself how advantageous it may be to the World, determined to conceal this Art no longer, but for the good of my Neighbor to publish it, by giving to CHYMISTS a perfect and fundamental information of this new-invented Art, that they may no longer for the future spend their Time and Money in long and tedious Operations, but may after a more easy way, by the help of my Furnaces, be able to effect many excellent things. Now this Book shall be divided into Five Parts, the first whereof shall teach how to build a Furnace, in which incombustible things are distilled and sublimed, and indeed such things which cannot be done by Retort or any other Vessels, and how the Spirits, flowers, and Oils of

Minerals, and Metals may by the help thereof be prepared, as also what their Use and Virtues are.

In the Second Part, shall be shown another Furnace, in which combustible things, as Vegetables, Animals, and Minerals are distilled and most perfectly subtilized: by help whereof many most excellent Medicaments for the cure of most grievous and otherwise incurable diseases may be prepared.

In the Third shall be taught a certain new invention hitherto unknown, of distilling Burning Spirits, as of Wine, Corn, Fruits, Herbs and Roots; as also the Waters of Vegetables and Animals, and that in a great quantity, in a short time, and without much cost; as also of boiling Beer, Mead, Wine, and other things, which otherwise are made in Copper or Iron Vessels; and all this by the help of Wooden Vessels, and benefit of small Copper, or Iron instruments of two or three pounds weight, and that after a certain easy manner without Furnaces. This newly-invented Art does also teach divers Chymical Operations, as Putrefactions, Digestions, Circulations, Extractions, Abstractions, Cohobations, Fixations, & etc. And this invention is very necessary and profitable for young beginners in this Art, for they need not in the making of burning Spirits, Waters of Vegetables, Extracts, and other Medicaments so many Furnaces, and so many Copper, Iron, Tin, Earthen and Glass Vessels, for it is here taught how all the aforesaid Operations may be done only by the help of a certain small Copper or Iron Instrument in Wooden Vessels as well as by Alembics and other great Copper Vessels, by which means a great deal of Cost is saved.

In the Fourth Part, shall be taught another certain, and hitherto unknown Furnace, in which all Chymical Operations may most easily be done: being most profitable for the trying of the Natures of Minerals and Metals; as also for the proving, examining, melting, cupelling, and separating of Metals, that nothing may be lost of them, and that after a compendious and easy way, and to great advantage.

In the Fifth shall be taught how to make and prepare Iron, Earthen, Glass and other kinds of Instruments necessary for the aforesaid four Furnaces, as also other necessary, and most profitable Manuals.

And in the First Part, the Fabric of the first Furnace being delineated, I shall also show how by the help thereof may be made Spirits, Oils, Flowers, and other profitable Medicaments, also their Virtues and Use, and that as faithfully as I may, and without fraud. And truly I do not doubt but those of understanding, will approve of this Work, but ignorant ZOILUS'S will condemn it: For it is said according to the Proverb, He that builds by the highway, will hear many things from them that find fault, and especially from the vulgar, & etc. But it would be well if those THRASOS would put forth something more excellent, before they find fault with and carp at other Men's pains and labors. Wherefore let no one rashly judge of this Work, until he be thoroughly informed concerning the same, and then I do not doubt but the Author shall be by him commended.

And if haply all things shall not presently succeed well, to his mind, with him that shall build this Furnace, and operate therewith, let him think with himself that perhaps he has erred in some part, (for it is a new and unknown work, in which any one may easily err) and not presently therefore murmur against the Author, blaming him, because he has not written clearly enough, but let him ascribe it to his own ignorance, and let him study to understand the Author's meaning, and still be practicing upon it, and then I do not doubt, but he will have better success, which I pray every one may have. AMEN.

THE COMPLETE WORKS
OF

RUDOLPH
GLAUBER

trans: Chris. Packe

RAMS
1983

PHILOSOPHICAL
FURNACES

FIRST PART

First Part of Philosophical Furnaces

Of the Structure of the First Furnace.

As for the first Furnace, it may be built greater or lesser as you please, a regard being had of the quantity of the Matter to be distilled, and either round or square; either of Bricks, or by a Potter with Potters Clay. Now when the Diameter is of one span, viz. with inside, the height must be of four, viz. one from the bottom to the grate, another from the grate to the bole made for putting in of Coals, and two from thence to the top of the Pipe, which must at least go forth out of the Furnace one span, lest the receivers should by the nearness of the Furnace be heated. The Pipe also must have on the fore part a Diameter, answering the third part of the intrinsic Diameter of the Furnace; also, a little larger on the hinder part than the forepart. Let the grate be such a one, as may be taken out at your pleasure and made clean, being stopped by the Matter that is cast in and distilled: for it is easily stopped in distilling of Salts melted with the coals, whereby the air is kept from coming to the fire, and the distillation by consequence hindered: Or let there be put into the Furnace cross-wise two strong iron bars, upon which lay four or five lesser, distant the one from the other the breadth of a finger, going a little out of the Furnace, by which when they are stopped, you may take them out with a pair of Tongs, and cleanse them from the burnt Matter, and then again put them into their own places: wherefore also the Furnace must on the fore part be open under the grate, that you may the better order the grate.

Also, the grate must have above it a covering of Iron or Stone, with a hole in the middle thereof with a certain distinction, which is to be filled with sand, that the cover may the better and more fitly shut the hole, and prevent the exhaling of the spirits which by this means will, being forced, go forth thru the Pipe into the receivers, after you have cast in the matter which is to be distilled.

Of the Receivers.

Let the Receivers be made of glass, or of strong earth, which may retain the spirits, and such is the Waldburgick, Hassiack, Frechheimensian, Siburgic earth, & etc. They are better that are made of glass, if they are to be had, and those especially which are made of strong and firm glass, which may be smoothed about the joints with a Smiris stone, and so fitted that they may the better be Joined together, and then they shall be smoothed with the Smiris stone, and be fitted, shall be taught in the Fifth part, which treats of Manuals, because by this means they are joined so close, that no spirits can go through the joints: otherwise you must close the joints with the best Lute, such as will not let the spirits exhale, which shall be taught in the Book treating of Manuals.

The form of the recipient you may see in the delineation thereof. As for the quantity thereof, know that by how much the greater they are, so much the better they are, for then you need the fewer, but the more, by how much the lesser they are. Let the superior orifice be larger than the inferior, so that always another receiver may with its inferior orifice be joined to it, and let the inferior orifice have a diameter of three fingers breadth, or thereabouts; I mean in case the Diameter of the Furnace be of one span. For a greater Furnace requires greater holes, as also orifices of the receivers, by which means a sufficient and due proportion of air may be given to the fire; or if the Diameter of the Furnace be more than a span, it must also have two or three pipes (which being considered together, should have a wideness answering the wideness of the third part of the Furnace, for so great a wideness, and so much air is required, if the fire burn freely and do its office) to which vessels of the aforesaid proportion must be applied, that the fire be not choked.

    Now, the Figure that is annexed will teach the conjunction of the Receivers, as also their application to the Furnace. And in the first place, the Receiver stands in a three-foot stool bored through in the middle, that the neck of the first Receiver may pass through, to which is applied a dish with a pipe receiving the dropping spirits: To the first there is joined a second, and to that a third, and so consequently (viz. near unto a wall or ladder) so many as you please. Let the upper Receiver, and indeed all the rest, be left open: To the lower as has been said, is Joined a dish with a pipe, by which the distilled Spirits run down into another certain glass vessel added thereunto, which being filled, is taken away, and another is set in the place of it, because that is set under it without luting, and therefore may easily be changed. And if you please to distil anything else, you may take away that dish with a pipe, and make it clean, and then Join it close again (that no spirit may breathe forth) to the neck of the lower receiver. And if that dish cannot be so closely joined, that nothing exhale, pour in a spoonful of Water, for that does astringe, neither does it hurt the spirits, because in the rectifying it is separated.

Of the Subliming Vessels.

These you need not make of glass, or of such earth as may retain the spirits, as has been above mentioned; it is sufficient, if so be they be made of good common Potters earth, and be well glazed within, viz. of such a form and figure, as appears by the annexed delineation.

Yet you must choose good earth that will endure the fire, for the lower pots are so heated by the fire, that they would be broken if they should not be made of good earth.

Now I will show you in general the manner itself of distilling; as also, the manual necessaries in every distillation.

The Manner of Distilling.

In the first place, let there be some burning Coals put in, which afterwards must be covered with more until the Furnace be full almost to the pipe, which being done, let not the uppermost cover be laid over its hole (that the heat and smoke may pass that way, and not thru the pipe, and receivers, which will thereby be red hot; and this will be a hindrance to the distillation) until the fire be sufficiently kindled, and the Furnace be thoroughly hot; then cast in, with an Iron ladle, of the matter prepared for distillation as much as will cover the Coals, which being done, Stop the Furnace very close, by pressing down strongly the upper cover upon its hole or sand, which is put in the lower part of the hole, being a place made for that purpose. Now let him that casts in anything thru the middle hole, presently stop it with a stopple of stone, and that very close, for by this means all those things which were cast in, will be forced, after the manner of a thick Cloud, to break forth through the pipe into the receivers, and there to condense themselves into an acid spirit or oil, and thence to distil into the dish set under, through the pipe whereof they do yet distil down further into another glass receiver. The Coals being burnt out, and all the spirits being come forth, you must cast in more Coals, and more materials, until you have got a sufficient quantity of Spirits. In this way of distilling, you may at your pleasure cease, and begin again without any danger.

    When you will make clean the Furnace, you need do nothing else, then draw out the Iron bars that lie on the cross bar, that the CAPUT MORTUUM may fall down, which afterwards may be taken away with a Fireshovel, which being done, you must put in the bars again, and lay them on the cross—bars as before, upon which you must cast burning Coals, and upon them others, until there be enough, then on them all, being well kindled, cast your materials.

    When you go to make clean the receivers, and to begin to distil another thing, you need not remove them, but only pour pure Water into them, viz. by their upper receiver, by the descending whereof the other are purified.

    And by this way, not only out of Vegetables, and volatile Minerals (incombustible) but also out of fixed Metals, and Stones, spirits, oils, and flowers, are drawn forth wonderfully, easily, and in good quantity, which otherwise could never have been done by the vulgar art of distilling.

Now, in this Furnace are distilled only such materials, which being distilled, yield an incombustible humidity, as common Salt, Vitriol, Allome, and other Minerals and Metals, each of which does yet require their peculiar manuals, if operated upon.

Now, because this Furnace does not serve for every matter, because the materials to be distilled are cast upon burning Coals, which are things combustible, I have determined in the second part to give another, viz. a lesser, unlike to this, yet convenient to distil all combustible things that are endued with volatile spirits, as Tartar, Harts-horn, Amber, Sal Armoniack, Urine, & etc. There are, by the help hereof, made most subtle, volatile, sulphureous spirits of Salts, and Minerals, as of common Salt, Vitriol, Allome, Nitre, Antimony, and of all other Minerals and Metals, which otherwise, without this Furnace, could not have been made, with which spirits, wonderful things are performed in Medicine and Alchymy, as in the Second Part shall be demonstrated more largely.

Now I will show you a way to make other Receivers belonging to the first Furnace, and indeed, such as are more fit for some Operations, as the former were more fit for others: wherefore let him that will operate, choose these, or the other, as he pleases.

As therefore the former being erected upwards by a wall, or ladder, by which means the spirit might ascend from one into another so long, until being refrigerated and condensed might again drop downward into the dish that is annexed thereto: so these are a contrary way set and placed collateral in a vessel with cold Water to condense the spirits, by which means you need not so many receivers; also they must not be fashioned like the former, as to be open above, and below, but only above like pots that serve for boiling: but this you must observe, that by how much the deeper and larger they are, by so much the better they are.

Also you must join them together by the help of earthen pipes, being so distinct, that the spirits may be kept back, being yet hot (and not refrigerated) from passing out of one into the other, but being forced through the middle of the separation of the pipes, may go to the bottom of every receiver, and thence arise by another pipe into another receiver that has a double cover like the former, where again descending to the cold bottom, remain refrigerated and condensed, Now three or four of these are enough (whereas of other, thirteen or fifteen are required) a regard being had of their Greatness.

You may see the figure of these receivers, as also their joining together by the annexed delineation, Now, for the most part, one is sufficient for him that distils a few things, especially if the matter be not precious, and then let one crooked earthen pipe at least be joined, one arm with the pipe that goes forth of the Furnace, the other with the Receiver, but so that it go into the receiver downwards, even to the middle thereof, and then you need not shut the orifice of the receivers, for it is no great matter if somewhat evaporate, viz. if the matter to be distilled be not precious. And by this way may new spirits and new flowers be made every hour, with the help of one Furnace, and one recipient, but with this caution, that for every new distillation, the recipient be washed with Water before it be put to the pipe; which being put to, you may then cast your species into the Furnace; and this do till you have a sufficient quantity of spirits.

And this way of distillation serves especially for the trying of the natures and properties of many and divers Minerals, such as yield in the fire spirits and flowers. For it would be too tedious to every new distillation to apply a new and distinct receiver: as also many studious of the Chymical art would quit their study, being able to make by retort but one trial in a day. And no wonder if expenses, and loss of time should deter many.

Now here there is no need of many Retorts, nor of luting them, nor of receivers, and such like superfluous things; neither is there here required the constant presence of the operator, the observation of the regiment of fire, the neglect whereof would otherwise endanger the loss of the retorts and receivers, and by consequence the loss of labor. These and such like tedious things are not here to be cared for, because it is sufficient only to cast the Matter upon the coals, and Cover the Furnace, and then presently go forth the spirits, and flowers of the same kind with their mineral: of which when you have a sufficient quantity, you must draw out the Iron bars, upon which the coals lye, that they may fall down, and be taken away; and whilst the Furnace is yet hot, to put in the Iron bars again, and upon them to lay fresh coals, which then will of their own accord be kindled with the heat of the Furnace. In the meantime, you must take away the receiver, and make it clean and set it to again, or if you had rather put another clean one, viz. for the new distillation of another Matter.

And by this way, divers things may be in the space of one hour distilled, and sublimed, viz. in a small quantity. But he that will distil, or sublime in a greater quantity, let him take three or four pots that the spirits may pass from one into another, that nothing thereof be lost. Here needs not (as I said before) the continual presence of the operator, for he may be gone, cease, or repeat as he please, because the work is without danger of breaking the retorts, and receivers.

He that knows the use of this Furnace, may do many things in a short time with little cost. For any one may do more by the help thereof in one hour, than in the common way in twenty-four, by which way also there is a great saving of coals, because ten pounds of coal will do more this way than a hundred the other. As for example, he that will try, shall make a pound of spirit of Salt in one hour with three, four, or five pounds of coal; whereas after the other way are required fifty or sixty pounds, and at least twenty or thirty hours' time, viz. in the common way by the help of retorts: which is indeed very tedious.

Also by this way may be made the flowers of minerals, and metals, in a great quantity, very easily, and in a short time without great cost, so as that in one hour's space, with three or four pounds of coal may a pound of the flowers of Antimony be made. And this is no small help to the Physician, and Chymist.

Moreover, this furnace being once built, endures for so many years, and being broken is easily repaired.

And by this way you shall need only materials to be distilled, no retorts and receivers are in danger, by which means much cost is saved.

Besides the aforesaid ways, I have yet another, and that more compendious, viz. of distilling, and subliming, and more easily, by which means in a very little time, an incredible quantity of spirits of Salts, and flowers of Minerals, and metals may be made; which I shall refer till another time, because for the present I have said enough.

Now I do not doubt, but diligent Chymists will follow my steps, and find out those things which are unknown to me. For it is easier to add to things found out, than to find out things unknown.

The construction therefore of the furnace being in my opinion clearly showed, there now follows the manner of distilling, and subliming with it.

Although haply, and contrary to my hope any obscurity should be met withal, yet one process will explain another; and the diligent operator, and searcher of Nature shall without doubt, by his practice attain the effect after the same manner as I have prescribed: which together with the blessing of GOD, I heartily wish all pious Chymists, Amen.

How the Spirit of Salt is to be distilled.

The reason why I enter upon the spirit of salt, before I say anything of the spirits of vegetables, is this, viz. because it is even the chiefest, which can be made in this furnace: for few exceed this in strength and virtues; wherefore I also have given it the preeminency. Neither is there any of the acid spirits, about which the Chymists hitherto have been more busied, than this, wherefore also it is of all, of greatest price, etc., for some have mixed salt with potter's clay, and have made this mixture into little balls, which they must get the spirit, forced by retort in a very strong fire: some have mixed salt with bole, some with the powder of tyles, others with burnt Allome, etc.

Others using a more compendious way have made salt to flow in a retort, which has a pipe both in the upper, and hinder part; by the upper pipe of which they have dropped in cold water, to elevate the ponderous spirits of the salt, but by the hinder they have blown with Bellows, to force the spirits into the retort: and this way is not altogether to be slighted, yet it has this inconvenience, that in process of time the retorts are broken that they can no longer retain the salt, and so the distillation is intercepted. Some have attempted it with Iron retorts, but by this means the spirits have been deaded, because they easily set upon the Iron, whence instead of spirit they have had flegme. And such, and tedious ways of distilling they have invented; and by the best of them indeed they could scarce distil one pound in 25, or 30 hours' space with 50, 60 or 100 pounds of coal; this being the reason, because the salt is very little wrought upon, and therefore it is that few ever had the spirit right and good, whence also the virtues thereof have been unknown.

And this therefore I was willing to make known, that it might appear, what price, this spirit has hitherto been of, and how easy, and abundantly, and with what little cost, it may after my new invented way be made.

It is said above, that the materials may in this way of distilling be immediately cast into the fire; yet this must be wisely understood. For although some of the species may without any preparation be immediately cast into the fire, yet it does not follow that all and every one of them must: for in some of them we must use our discretion, as in the distilling of salt. For if the salt be immediately cast into the fire, it will not only yield no spirits, but will leap so long upon the coals, until it finds a descent to the lowest part of the furnace:

Now this may be prevented divers ways; and first indeed after this manner: Dissolve salt in common water, then quench burning coals with this water, that they may be impregnated with the salt, which afterwards set on fire in the furnace: but you must first cast in other burning coals, upon which you must cast those that are impregnated with salt until the furnace be full, as is above said: and while the coals burn, the salt is resolved by the force of the fire into spirit.

Now you must observe that he that distils spirit of salt after this manner, must make choice of glass receivers, because the spirit whilst it is hot, penetrates because of its wonderful subtlety, those that are earthen. And this spirit is of a most grateful taste. But in defect of glass receivers, I shall show you another way wherein you may use those that be of earth.

Mix salt, and vitriol or allome together, grinding them very well in a Mortar (for by how much the better they are ground, the more Spirit they yield). Then cast this mixture into the fire with an Iron Ladle, viz. so much of it as will be sufficient to cover the coals, and then with a great fire the spirits come forth into the receivers, where being coagulated, they distil down into the dish, and thence into another receiver. And if you know how to work aright, the spirits will like water continually run out thru the pipe, the thickness of a straw; and you may easily every hour make a pound of the spirit. Now the reason why you shall by this way have more spirits than by the other, is this, viz. because the vitriol and allome, which is mixed with the salt, makes it flow quickly, by which means it is prevented from falling down through the coals to the lower part of the furnace, but sticking to the coals is almost all of it turned into spirits. The CAPUT MORTUUM, which is reddish, easily falls with the ashes through the grate, and can no more be distilled, but yields by excoction a white fixed salt, which serves for the flowing of metals; and being dissolved in warm water serves also for a glyster against the Worms, which it kills, and purges also the Bowels.

You might object, that the spirit made after this manner, is not the true spirit of salt because of the mixture of vitriol and allome, but mixed, and compounded. I answer: There can by this way distil no spirit of vitriol, or allome, being that which I often tried, casting vitriol or allome into the furnace, where I received no spirit at all, the reason of this is, because these spirits are far more heavy than the spirit of salt, neither can they ascend so great a height, viz. of three spans, but are burnt, whence unless the flegme, nothing distils. Wherefore the spirit of salt that is made after this manner is not mixed, but pure and mere spirit of salt, of the same taste and virtue as that is of, that is made by itself; because in this furnace the spirit of allome and vitriol, cannot be made unless a pipe go out of the furnace near the grate, as you may see by the delineation of the furnace, for otherwise it cannot be made; besides, these spirits are better, and more truly taught in the second part. And if it be granted that somewhat together with the spirit of salt comes forth (which is yet impossible) what hurt I pray you comes from thence either in the solution of metals, or medicine? Wherefore the spirit made after this way is not to be suspected. Yet I will satisfy the incredulous, and will show him another way without the addition of allome or vitriol, for the distilling of that spirit, but that will be in the second part of this Book, where I will teach you the furnace, by which is made spirit of Nitre, Aquafortis, and amongst combustibles, the Oils of vegetables, and Fats of animals and other things which cannot be made by this: and by this way I will satisfy those, who are not pleased with the former.

Now for the want of glass receivers, we are forced to use earthen, but these cannot retain the spirit of salt made after the aforesaid ways; in which case I could indeed discover a certain little manual, by which the aforesaid spirit may be received even in a great quantity in earthen recipients: but for certain causes I shall here be silent, and shall refer it till the edition of the second part. Let it suffice therefore that I mentioned such a thing, wherefore omitting that, I shall proceed to show you the virtues, and use of this spirit, as well in Alchymy, as in Medicine, and other Mechanical Arts.

Of the Use of the Spirit of Salt.

It is worthwhile, to speak of the power, and virtues of this excellent spirit; what other Authors have clearly described, I shall here pass over, and refer the Reader to the writings of those Authors; touching only on some few of which they said nothing.

The Spirit of salt is by most accounted a most excellent medicine, and safely to be used, as well inwardly as outwardly: it extinguishes a preternatural thirst in hot diseases, absterges and consumes flegmatick humours in the Stomach, excites the Appetite, is good for those who are hydropical, have the Stone, and Gout, & etc. It is a menstruum dissolving metals, excelling all other therein: For it dissolves all metals and minerals (excepting silver) and almost all stones (being rightly prepared) and reduces them into excellent medicaments. It does also many excellent things in mechanical arts.

    Neither is it to be slighted in the kitchen, for with the help thereof are prepared divers pleasant meats for the sick as well as for those that are in health, yea and better than with Vinegar, and other acid things: and it does more in a small quantity, than Vinegar in a great. But especially it serves for those Countries that have no Vinegar. It is used also instead of Verjuice, and the juice of Lemons. For being prepared after this way, it is bought at a cheaper rate than Vinegar or juice of Lemons. Neither is it corruptible as expressed juices are, but is bettered by age. Being mixed with Sugar it is an excellent sauce for roast meat. It preserves also divers kinds of Fruits for many years. It makes also Raisons, and dried Grapes to swell, to acquire their former magnitude again, which are good to refresh a weak stomach in many diseases, and serves for the preparing of divers kinds of meats of Flesh and Fish; but you must mix some water with the spirit, or else the Raisons will contract too much acidity. This spirit does especially serve for making meats delightfully acid; for whatsoever things are prepared with it, as Chickens, Pigeons, Veal, & etc. are of a more pleasant taste than those which are prepared with Vinegar. Beef being macerated with it, becomes in a few days so tender, as if it had been long time macerated with Vinegar. Such, and many more things can the Spirit of Salt do.

A distillation of Vegetable Oils, whereby a greater quantity is acquired, than by that common way, by a Vesica.

As many Distillers, as hitherto have been, have been ignorant of a better way to distil Oils of Spices, Woods, and Seeds, than by a vesica or alembic, with a great quantity of water. And although they may also be made by retort, yet there is a great deal of care required, or else they contract an EMPYREUMA, wherefore that way, by a still, is always accounted the better, which way indeed is not to be slighted, if you distil Vegetables of a low price, and such as be oleaginous; but not so in the distillation of Spices, and of other things that are of a greater value, as are Cinnamon, Mace, Saffron, & etc. which cannot be distilled in a gourd still without loss, because then there is required a great quantity of water, and by consequence great, and large vessels, to which something adheres, wherefore we lose almost half, which is not to be so much valued in vegetables that are oleaginous, as in Anise seed, Fennel, and Carroway seed, & etc. But the loss made in distilling of drier and dearer vegetables, as Cinnamon, LIGNUM RHODIS, CASSIA, is evident enough, and by consequence not to be slighted. Neither can it be distilled that way, for a good quantity by coction acquires a gummy tenaciousness, which cannot ascend with the water. But that this way for the future may be prevented, I will show another way to distil the Oils of Spices, and other precious things, which is done with Spirit of Salt, whereby all the Oil is drawn forth without any loss, the process whereof is this, viz. Fill a gourd with Cinnamon or any other Wood, or Seed, upon pour so much of the spirit of salt, as will be sufficient to cover the wood, then place it with its Alembic in Sand, and give it fire by degrees that the spirit of salt may boil, and all the Oil will distil off with a little flegme; for the spirit of salt does with its acrimony penetrate the wood, and frees the Oil that it may distil off the better and easier. And by this way the Oil is not lost by the addition of that great quantity of water in those great and large vessels, but is drawn in lesser glass vessels with the addition of a little moisture. Distillation being finished the spirit is poured off by inclination from the wood, being again useful for the same work. And if it has contracted any impurity from the wood, it may be rectified: but residue of the spirit which remains in the wood you may recover, if that wood is cast into the aforesaid furnace upon burning coals, by which means it may come forth again pure, and clear: and by this means we lose none of the spirit of Salt. And after this way by help of the spirit of Salt, are drawn forth Oils of dearer Vegetables together with their Fruit, which cannot be done by a still.

There are made also by means thereof of Oils of Gums and Rosins, clear, and perspicuous.

The clear Oil of Mastick, and Frankincense.

Take of Frankincense or Mastick powdered small, as much as will serve to fill the third part of a Retort (which must be coated) upon which pour a sufficient quantity of spirit of Salt, taking heed that the Retort be not filled too full, or else the spirit when it boils, flows over it, then place it in sand, and give fire by degrees, and there will first come out some phlegme, after which a clear transparent oil together with the spirit of salt, which must be kept by itself, after this a certain yellow Oil which must be received by itself, and last of all there follows a red Oil, which although it is not to be cast away, yet it is very unlike to the first, serving for outward uses, and to be mixed with Ointments and Emplasters, for it does wonderfully consolidate, and therefore good in new and old Wounds. The first being well rectified, is in its subtlety, and penetrating faculty not unlike to spirit of wine, and may profitably be used inwardly and outwardly, viz. in cold effects, but especially in the stiffness of the Nerves, caused by cold humours, upon which follows a contraction; but then you must first rub the member contracted with a linen cloth, that it may be well warmed, into which then the Oil must be chafed with a warm hand. For it does do wonders in such like effects of the Nerves.

    After the same manner, may Oils be made from all gums. The red, tenacious and stinking Oils of Tartar, Harts-horn, Amber, & etc. distilled after the common way by retort are also rectified with spirit of salt so as to become transparent and to lose the EMPYREUMA contracted by distillation,

    Now the cause of the blackness, and fetidness of these kind of Oils, is a certain volatile salt which is to be found as well in Vegetables, as certain Animals, which is easily mixed with the Oil, and makes it of a brown color. For every volatile salt whether it be of Urine, Tartar, Amber, Harts-horn, and of other Vegetables and Animals, is of this condition and nature, as to exalt, and alter the colors of sulphureous things, and that either for the worse or for the better: but for the most part it makes Oils thick, black and stinking, as you may see in Amber, Harts—horn, and Tartar. The cause therefore of the blackness, and fetidness of these Oils being known we may the more easily take heed thereof in distilling, and being contracted, correct them again by the help of spirit of Salt. For all volatile Salt has contrariety to any acid spirit, and on the other side, every acid spirit has a contrariety with all volatile salts, that have the nature of salt of Tartar. For metals that are dissolved with acid spirits are as well precipitated with spirit of Urine, or any volatile salt as with the liquor of salt of Tartar; which shall be more at large declared in the second part.

The volatile salt therefore is by the mortifying acid spirits, as of Salt, Vitriol, Allome, Vinegar, & etc., deprived of its volatility, and is fixed, by which means being debilitated it forsakes its associate which was infected with blackness by it: it is necessary that we should proceed after the same manner with these fetid Oils, viz. as follows.

Take any fetid Oil of Tartar, Amber, & etc., with which fill the fourth part only of a glass Retort, and upon it pour by drops the spirit of salt; and it will begin to be hot, as it is used to be, when Aqua fortis is poured on salt of Tartar; wherefore the spirit is to be poured on it by little and little, and by drops for fear of breaking the glass.

Now the sign of the mortification of the volatile salt is, when it ceases to make a noise, and then no more is to be poured on, but set your Retort in sand, & give fire to it by degrees, as is used to be done in the rectifying things of easy elevation: and first of all will go forth a certain stinking water, after which comes a transparent clear, and odoriferous Oil, and after that a certain yellow, clear, and also well smelling Oil, but not so as the first, wherefore each must be taken apart by changing the receivers.[6] Now these Oils become more grateful than those fetid ones of the shops. For these Oils retain their clearness, and fairness, the cause of their fetidness, and redness being taken away by the spirit of salt. In the bottom of the retort remains the black volatile salt with the spirit of salt, from whence it may be sublimed into an odoriferous salt resembling salt armoniack in taste. The spirit of salt is also deprived of its acidity, and coagulated by the volatile salt, and is like TARTARUM VITRIOLATUM, appointed also for its uses, as shall be spoken in the second part, of the spirit of Urine.

After the same manner, also are rectified other Oils, which by length of time have contracted a clamminess, as are Oil of Cinnamon, Mace, Cloves, & etc. with the spirit of Salt, if they be rectified by Retort, for then they acquire again both the same clearness, and goodness, as they had when they were newly distilled.

Here I must make mention of a certain error of Physicians, not only of ignorant Galenists but Spagyricks, committed in the preparations of some Chymical medicaments. For many have persuaded themselves that Oil of Tartar, Harts-horn, & etc. having lost its stink, is a Medicine radically taking away all obstructions; but this must be taken with a grain of salt. For some have rectified these kinds of Oils by calcining Vitriol, and by that means have somewhat made them lose their EMPYREAMA,

---

[6] This is an example of fractional distillation. -PNW

but with all their Virtues; which others observing have conceived that the fetidness thereof is not to be taken away, because the Virtue of them is thereby lost, as if the Virtue consisted in the fetidness thereof; but that is a very great error, because fetidness is an enemy to the heart and brain, and in it is no good. But this is granted, that they that take away the fetidness of those Oils mortify the virtues of them. But you might say, how then must we proceed in taking away their fetidness without the loss of the virtues? Must they be rectified by the spirit of salt? As even now you taught. Response: No, for although I said that Oils might be clarified with spirit of salt, yet it does not follow that my meaning was, that that clarification was the mending of them: This is only a way of clarification, whereby they become more grateful; and it is not to be slighted, a better being unknown. But how they are to be rectified from their fetidness and blackness, without the loss of their Virtues, and to be made more noble, does not belong to this place, because it cannot be done by this Furnace: I shall refer the reader therefore to the second part, where it shall be showed, how such spirits are to be rectified without the loss of their virtues, which being so prepared may well be accounted for the fourth Pillar of Physick. And these things I was willing at least for information sake to show you, not to offend you, and that because I was moved with pity, and compassion towards my neighbor.

The Quintessence of all Vegetables.

Pour upon Spices, Seeds, Woods, Roots, fruits, Flowers, & etc. the Spirit of Wine well rectified, place them in digestion to be extracted, until all the essence be extracted, with the Spirit of Wine; then upon this Spirit of Wine, being impregnated, pour the best Spirit of Salt; and being thus mixed together, place them in Balneo to digest, until the Oil be separated, and swim above from the Spirit of Wine, then separate it with a separating glass, or distil off the Spirit of Wine in Balneo, and a clear Oil will ascend; for if the Spirit of Wine be not abstracted, then that Oil will be as red as blood; and it is the true quintessence of that vegetable, from whence by the Spirit of Wine it was extracted.

The Quintessence of all Metals and Minerals,

Dissolve any metal (excepting Silver, which must be dissolved in Aqua fortis) in the strongest spirit of Salt, and draw off the flegme in Balneo; to that which remains pour the best rectified spirit of Wine, put it to digesting, until the Oil be elevated to the top as red as blood, which is the tincture, and quintessence of that metal, being a most Precious treasure in medicine, a sweet and red Oil, of Metals and Minerals.

Dissolve a Metal or Mineral in spirit of Salt, dissolve also an equal weight of salt of Wine essentificated; mix these dissolutions, and distil them by retort in a gradual heat, and there will come out an oil sweet, and as red as blood, together with the spirit of Salt; and sometimes the neck of the retort and receiver will be colored like a Peacocks tail with divers colors, and sometimes with a golden color.

And because I would without any difference comprehend all Metals and Minerals under one certain general process; let him that would make the essence of silver take the spirit of nitre, and proceed in all things as was spoken of the other metals. Concerning the use of these essences, I need not speak much hereof; for to him that knows the preparation shall be discovered the use thereof. Concerning the corrosive oils of metals and minerals, seeing they cannot be described by any one process, it will be worthwhile to set down what is peculiar to each of them, as follows.

The Oil, or Liquor of Gold.

Dissolve the calx of gold in the spirit of salt, (which must be very strong, or else it cannot dissolve it) but in defect of the strongest spirit thereof, mix a little of the purest salt—peter; but that oil is the best which is made with the spirit of salt alone. From the gold dissolved abstract half the solution, and there will remain a corrosive oil, upon which pour the expressed juice of lemons, and the dissolution will become green, and a few feces fall to the bottom, which may be reduced in melting. This being done, put this green liquor in Balneo, and draw off the flegme: that which remains take out, and put upon a marble in a cold moist place, and it will be resolved into a red oil, which may safely, and without danger be taken inwardly, curing those that are hurt with Mercury. But especially it is commended in old ulcers of the mouth, tongue, and throat, arising from the French pox, leprosy, scorbute, & etc. where the oil of other things cannot be so safely used. There is not a better medicine in the exulceratlon, and swelling of the glandules, in the ulcers of tongue and jaws, which does sooner mundify, and consolidate. Neither yet must we neglect necessary purgings, and sudoroficks, for fear of a relapse, the cause not being taken away.

Neither will there any danger follow, whether it be given inwardly, or used outwardly, as in the accustomed use of other medicaments, and gargarisms; for it may daily, and truly without all danger be used at least three times with a wonderful admiration of a quick operation.

Oil of Mars.

Dissolve thin plates of Iron in rectified spirit of salt, take the solution, which is green, of a sweet taste, and smelling like fetid sulphur; and filter it from that filthy and feculent residence; then in a glass gourd in sand, abstract all the humidity (viz. with a gentle fire) which will be as insipid as rain-water, because the iron by reason of its dryness, has attracted all the acidity to itself: but in the bottom will remain a mass as red as blood, burning the tongue like fire: it takes away all proud flesh of wounds, and that without danger. It is to be kept in a glass close stopped from the air, lest it be resolved into an oil, which will be of a yellow color. But he that desires to have the oil, may set it on a marble in a moist Cellar, and within a day it will be resolved into an oil, which will be in color betwixt yellow and red: It is a most excellent secret in all corroding ulcers, fistulas, cancer, & etc., being an incomparable consolidator, and mundifyer. And it is not without profit mixed also with common water to wash the moist, fetid ulcers of the legs; which cause tumors, by being applied warm like a bath, for it dries, and heals suddenly, if withal Purges be administered it cures also any scab. That red mass (being yet unresolved) being put on the oil of sand, or flints (of which in the second part) makes a tree to grow in the space of one or two hours, having root, trunk, and boughs: which being taken out, and dried, in the test yields good gold, which that tree extracts from the earth, i.e. from the flints, or sand. You may, if you please, more accurately examine this matter.

Oil of Venus.

Spirit of Salt does not easily work upon Copper, unless it be first reduced into a calx, and that after this manner. Take plates of Copper made red hot in an open crucible, quench them in cold water, and they will cleave into red scales: then the remainders of the plates make red hot, and quench as before: do this so often, till you have obtained a sufficient quantity of the calx; which being dried, and powdered, extract with the rectified spirit of salt, in sand, until the spirit of salt be sufficiently colored with a green tincture, which you must decant, and filter; and then abstract from it the superfluous moisture, that there may remain a green thick oil, which is an excellent remedy for ulcers, especially such as are Venereal, being applied outwardly.

Oil of Jupiter and Saturn.

Neither are these two metals easily dissolved in the spirit of salt, yet being filed, are dissolved in the best rectified spirit of salt. But the operation is performed better with the flowers of these metals (the preparation whereof shall be hereafter taught.) Take therefore the flowers, upon which in a gourd glass pour the spirit of salt, and presently the spirit will work upon them, especially being set in a warm place; filter the yellow solution, and abstract the humidity, until there remains a yellow heavy oil, which is proper against putrid ulcers.

Oil of Mercury.

Neither is this easily dissolved with the spirit of salt: but being sublimed with vitriol, and salt is easily dissolved. Being dissolved, it yields an oil very corrosive, which must be used with discretion, wherefore it is not to be administered, unless it be where none of the other are to be had. For I saw a woman suddenly killed with this oil, being applied by a certain Chyrurgeon. But this oil is not to be slighted in eating ulcers, tetter, & etc. which are mortified by it.

Oil of Antimony.

Crude Antimony that has never undergone the fire, is hardly dissolved in spirit of salt: as also the REGULUS thereof; but the REGULUS being subtly powdered, is more easily wrought upon, in case the Spirit be sufficiently rectified.

The VITRUM is more easily, but most easily of all the flowers are dissolved, being such as are made after our prescription a little after set down. Neither is BUTYRUM ANTIMONII (being made out of sublimed Mercury, and Antimony) anything else but the REGULUS of Antimony dissolved with spirit of salt; for sublimed Mercury being mixed with Antimony, feeling the heat of the fire, is forsaken by the corrosive spirits associating themselves with the Antimony, whence comes the thick Oil; while this is done the sulphur of Antimony is joined to the Quicksilver, and yields a Cinnabar, sticking to the neck of the Retort; but the residue of the Mercury remains in the bottom with the CAPUT MORTUUM, because a little part thereof does distill off: And if you have skill you may recover the whole weight of the Mercury again.

And these things I was willing the rather to show you, because many think this is the Oil of Mercury, and therefore that white powder made thence by the pouring on of abundance of water they call MERCURIUS VITA, with which there is no mixture at all of Mercury, for it is mere Regulus of Antimony dissolved with spirit of Salt, which is again separated, when the water is poured on the Antimonial butter; as is seen by experience; For that white powder being dried, and melted in a crucible yields partly a yellow glass, and partly also a Regulus, but no Mercury at all.

Whence it does necessarily follow that that thick oil is nothing else but Antimony dissolved in spirit of Salt. For the flowers of Antimony being mixed with spirit of Salt, make an oil in all respects like to that butter which is made of Antimony, and sublimated Mercury, which also is after the same manner by the affusion of a good quantity of precipitated into a white powder, which is commonly called MERCURIUS VITA: It is also by the same way turned into BEZOARDICUM MINERAL, viz. by abstracting the spirit of Nitre, and it is nothing else but Diaphoretick Antimony.

For it is all one whether that Diaphoretick be made with spirit of Nitre, or with Nitre itself, viz. corporeal, for these have the same virtues, although some are of opinion that that is to be preferred before the other; but the truth is, there is no difference. But let everyone be free in his own judgement, for those things which I have wrote, I have not Writ out of ambition, but to find out the truth.

Now again to our purpose, which is to show an oil of Antimony made with the spirit of salt.

Take a pound of the flowers of Antimony (of which a little after) upon which pour two pounds of the best rectified spirit, mix them well together in a glass, and set them in sand a day and night to dissolve, then pour out that solution together with the flowers into a retort that is coated, which set in sand, and first give a gentle fire, until the flegme comes off, then follows a weak spirit with a little stronger fire, for the stronger spirits remain in the bottom with the Antimony: then give a stronger fire, and there will come forth an oil like to the butter of Antimony made with sublimed Mercury, and is appropriated to the same uses, as follows.

The flowers of Antimony, White and Vomitive.

Take of this butter as much as you please, upon which in a glass gourd, or any other large glass pour a great quantity of water until the white flowers will precipitate no more; then decant off the water from the flowers, which edulcorate with warm water, and dry with a gentle heat, and you shall have a white powder.

The Dose is, that 1, 2, 3, 8, or 10 grains be macerated for the space of a night in wine, which is to be drank in the morning, and it works upward and downward. But it is not to be given to children, those that be old, and weak, but to those that be strong, and accustomed to vomiting. When at any time this infusion is taken and does not work, as sometimes it falls out, but makes the Patient very sick, he must provoke vomiting with his finger, or else it will not work, but make those that have taken it to be sick, and debilitated even to death. We must also in the over much working of these flowers drink a draught of warm Beer, or rather of warm Water, decocted with Chervil, or Parsley, and they will work more mildly. But let not him that is able to bear the operation thereof any way hinder it, for there is the greater hope of recovering his health thereby, for they do excellently purge choler, and evacuate flegme in the stomach, being humours that will not yield to other Cathartic; they open obstructions, resist the putrefaction of the blood, the causes of many diseases, such as are Fevers, Headaches, etc. they are good for them that are Leprous, Scorbutical, Melancholical, Hypochondriacal, infected with the French Pox, and in the beginning of the Plague. In brief, they do work gallantly, and do many things.

After the taking of them, the Patient must stay in his bed or at least not go forth of his house, for to avoid the air, or otherwise they may be mistrusted.

And because of their violence they are feared, and bated, I shall in the fourth part of this Book for the sake of the sick set down such as are milder, and safer, such as shall work rather downward than upward, causing easy vomits, which also you may give to children, and those that are old without danger, yet some respect being had of the disease, and age.

The Flowers of Antimony Diaphoretical.

The aforesaid flowers if they be cast into melted Nitre, and be left a while in melting, are made fixed, to become Diaphoretical, and lose their Cathartical Virtue. The acid water being separated from the flowers, if it be evaporated, leaves behind the best spirit of salt, serving for the same or such like uses again.

Of the External use of the Corrosive Oil of Antimony.

This oil has been long used by Chirurgions, for they have with a feather applied it to wounds almost incurable, to separate impurities, for the acceleration of the cure, that afterwards other medicaments being applied may the better operate. But it is better if it be mixed with spirit of Salt, for they are easily mixed and it is made more mild thereby, and the too great corrosive faculty thereof is mitigated. Neither is there any other besides the spirit of Salt, with which this oil can be mixed, unless it be the strongest spirit of Nitre, for the weak spirit of Nitre precipitates the butter of Antimony, as you may see in the preparation of BEZOARDICUM MINERALE. But the strongest spirit of nitre dissolving this butter, makes a red solution of wonderful Virtue in Chymistry, of which we are not to treat in this place; and if this be drawn off again by distillation, it leaves behind the first time a fixed Antimony, and Diaphoretical, which otherwise must be drawn off twice, or thrice, viz., if it be weak, and not able to dissolve the butter without precipitation.

Now this BEZOARDICUM is the best, and safest Diaphoretick in all diseases that require sweat, as in the plague, French pox, fevers, scorbute, leprosy, & etc. if it be given from 6, 8 to twenty grains in proper vehicles; it penetrates the whole body, and evacuates all evil humours by sweat and urine.

The Oil of Arsenic and Auripigmentum.

As the spirit of salt does not easily work upon Antimony because of the abundance of crude sulphur, unless it be reduced into flowers, in the preparation whereof, some part of its sulphur is burnt; so also, ARSENIC and AURIPIGMENTUM are hardly dissolved with spirit of salt, unless they be reduced into flowers, and the spirit of salt be very strong, which may be able to work upon it. These may be distilled by retort like Antimony into a thick heavy oil; which being used in cancerous eating ulcers, exceeds that of Antimony in mortifying, mundifying, and purging those evils. After the same manner, may corrosive oils be made out of all the realgars being ordained for outward uses.

Oil of Lapis Calaminaris.

Take of the best yellow or red LAPIS CALAMINARIS very subtly powdered, as much as you please, and pour upon it five or six times as much of rectified spirit of salt, mix and stir them well together, and do not leave them long unstirred, but ever and anon shake the glass with the materials; and this do oftentimes, or else the LAPIS CALAMINARIS will grow together into a very hard stone, which can be dissolved no more, and is prevented by the aforesaid often shaking: and when the spirit of salt will dissolve no more thereof in FRIGIDO, set the glass in warm sand so long, until the spirit be tinged with a most yellow color, which then decant, and pour on fresh, and again set it in digestion to extract, and do not to forget to shake the glass often. The solution being finished filter it, and cast away the residue of the TERRA MORTUA. Afterwards set the solution in sand, and give fire, and almost three parts of the spirit of salt will go over insipid, which is nothing but the flegme, although the spirit was never so well rectified; the reason whereof is the most dry nature of LAPIS CALAMINARIS, to which the spirit of salt is very friendly, and therefore very hard to be separated from it. For I never knew any mineral or metal (besides ZINK) which exceeds LAPIS CALAMINARIS in dryness. At last when no more flegme will go over, let all things cool; which being done, take out the glass, and you shall find a red thick oil, as fat as oil olive, and not very corrosive; for that spirit of salt being almost mortified with the LAPIS CALAMINARIS is deprived of its acidity. This oil is to be kept from the air; or else within a few days it attracts much air which it converts into water, and thereby becomes weakened.

    This Oil is of wonderful Virtue, being used as well inwardly as outwardly. And I wonder that in so long a time there has been nobody, who has operated in LAPIS CALAMINARIS and described the nature thereof, seeing it has in it a golden sulphur (of which thing in the fourth part) for if the terrestreity thereof were separated from it artificially, pure gold would be manifested therein; now the greatest part thereof is volatile, and immature, and cannot easily be reduced into a body in melting, wherefore hitherto that stone has not been esteemed of by Chymists, but to the wise was always precious, & etc.

The Use of the Oil of LAPIS CALAMINARIS,

If it be given from 1, 2, 3 drops to ten, and indeed with suitable vehicles, it purges the dropsy, leprosy, gout, and other noxious fixed humours not yielding to vegetable Cathartics, of which more at large in the Second Part of the spirit of urine, and salt of tartar. It serves outwardly for an excellent vulnerary balsome, the like to which can scarce be showed, not only in reducing old corrupt wounds, but also in those that are green, for it does powerfully dry, mundify, and consolidate.

It is also used in household affairs, for birdlime being dissolved in it, yields a certain tenacious matter serving to catch birds, mice, & etc. about the house or in the field. For it is as permanent in the heat of the Sun, as in the cold of Winter, wherefore it may be used at any time of the year, all small animals stick to it if they do but touch the matter.

A ligature of string smeared therewith, and bound about any tree prevents the spiders from climbing up thereon, and other kinds of insects that are noxious to the fruit; a thing worth taking notice of.

This oil is not by the pouring on of water corrupted, neither is it precipitated, as that of Antimony: wherefore it is useful for many things. Common yellow sulphur boiled in it, viz. in a strong fire, to be dissolved in it, swims upon it like fat, is thereby purified and made as transparent as yellow pellucid glass, and a better medicine then those common flowers of sulphur: it serves also for other uses, all which to relate here it would be too tedious.

This oil being mixed with clean sand, and distilled by retort in a fire that is very strong (otherwise the spirit of salt will not leave the LAPIS CALAMINARIS) yields a most fiery spirit, the LAPIS CALAMINARIS remaining in the bottom of the retort.

This spirit is so strong, that it can scarce be kept, it dissolves all metals, and all minerals (excepting silver and sulphur) wherefore by the help thereof many excellent medicaments are made, which cannot be made with the common spirit though never so well rectified, which although it be often rectified, yet it is not without flegme, which cannot be separated from it by the power of rectification, so well as with LAPIS CALAMINARIS.

This spirit does perform many things in medicine, & alchemy, as also in other arts, as you may easily conjecture; but here is not opportunity to speak more of these things, yet for the sake of the sick I shall add one thing to which few things are to be compared; the plain & short process whereof I would not have you be offended at. And it is this, viz. mix this spirit with the best rectified spirit of wine, digest this mixture some while, and the spirit of salt will separate the spirit of wine, and will make the oil of wine swim on the top, the volatile salt being mortified: and this oil is a most incomparable cordial, especially if with the said spirit of wine, spices have first been extracted, and with the said spirit of salt, gold has been dissolved. For then in the digestion of this mixture, the oil of wine being separated, attracts the essence of the cordial species, and of other vegetables, being extracted before with the spirit of wine, as also the tincture of gold, and so by consequence a most efficacious incomparable and universal medicine for all diseases, fortifying the HUMIDUM RADICALE, that it may be able to overcome its enemies; for which let praise and glory be given to the immortal God for ever who has revealed to us so great secrets.

Of the Extrinsical Use of the Spirit of Salt in the Kitchen.

I said before that instead of Vinegar, and verjuice it may be used, as also instead of the juice of Lemons, now it remains that I show you how it is to be used, and that indeed as well for the sake of the healthy as the sick.

Let him therefore that will dress a pullet, pigeons, veal, & etc. in the first place put a sufficient quantity of spices, of water, and butter, and then as he pleases a greater, or lesser quantity of spirit of salt: and by this means fleshes are sooner made ready being boiled, then that common way; an old hen though the flesh thereof be old is made as tender as a chicken by the addition of this spirit: but be that will use it instead of the juice of Lemons with roast meat, must put into it the pill of Lemons for preservation sake, because it preserves it. It is used instead of verjuice by itself alone, or mixed with a little sugar, if it be too acid.

He that will stew beef, and make it as tender as kid, must first dissolve it in tartar and a little salt before he wets the flesh therewith, and the flesh will not only be preserved but made tender thereby: but to keep flesh a long time you must mix some water therewith, and with weights press down the flesh, that it may be covered with the pickle: for by this means flesh may be preserved a great while.

After the same manner, may all kinds of garden fruits be preserved, as cucumbers, purslane, fennel, broom, German capers, & etc. and indeed better than in vinegar. Also, flowers, and herbs may a long while be preserved by the help thereof, so that you may have a rose all the winter.

It preserves also wine, if a little be mixed therewith. A little thereof being mixed with milk precipitates the cheese, which if it be rightly made is never corrupted, being like to such cheese as they call PARMESAN. The whey of that milk dissolves Iron, and cures any scab being washed therewith.

With the help of spirit of salt is made with honey, and sugar a most pleasant drink, not unlike to wine. There is made also of certain fruits with the spirit of salt a very good vinegar like to the Rhenish vinegar. Such and many more things, which I will not now divulge, may be done with spirit of salt.

And thus, have I in some measure taught the use of the spirit of salt, which I would not have you take as if I had revealed all things; for, brevities sake, as also some other reasons I have silently passed over many things. Neither do I know all things myself: but those things, which I do know, I have so far declared that others may from hence have hints of seeking further. He that would describe all, and every power and virtue thereof, had need to write a whole volume, the which is not my purpose at this time to do, but may perhaps be done another time. There shall also be showed in the second part of this book, some secrets which may be prepared by the help of this spirit: as how it may be dulcified to extract the tincture of gold, and of other metals, leaving a white body, which tincture is a medicine not to be slighted. Wherefore now seeing it is manifest how great things this Spirit can do, everyone will desire a good quantity for his household uses, especially seeing most excellent spirits may be made after an easy and short way.

How an Acid Spirit, or Vinegar may be Distilled out of all Vegetables, as herbs, woods, roots, seeds, & etc.

First put a few living coals into the furnace, then put upon them the wood that is to be distilled, that it may be burnt: out of which, whilst it is burning goes forth the acid spirit thereof into the receiver, where being condensed it falls into another receiver, resembling almost common vinegar in its smell, wherefore also it is called the VINEGAR OF WOODS.

And after this manner you may draw forth an acid spirit out of any wood, or vegetable, and that in a great quantity without Costs, because the wood to be distilled is put upon a very few living coals, and upon that another, for one kindles the other: and this spirit requires no more charges than of the wood to be distilled; which is a great difference betwixt this, and the common way of distilling, where besides retorts, is required another fire; and out of a great retort scarce a pound of spirit is drawn in the space of five or six hours, whereas in ours in the space of one day, and that without any cost or labor may be extracted twenty or thirty pound, because the wood is immediately to be cast into the fire to be distilled, and that not in pieces, but whole. Now this spirit (being rectified) may commodiously be used in divers Chymical operations, for it does easily dissolve animal stones, as the eyes of Crabs, the stones of Perches, and Carps, Corals also and Pearl, & etc. as does vinegar of wine. By means thereof also are dissolved the glasses of metals, as of tin, lead, Antimony, and are extracted, and reduced into sweet oils.

This vinegar being taken inwardly of itself does cause sweat wonderfully, wherefore it is good in many diseases, especially that which is made of Oak, Box, Guaiacum, Juniper, and other heavy woods; for by how much the heaver the woods are, by so much more acid spirit do they yield.

Being used outwardly it mundifies ulcers, wounds, consolidates, extinguishes, and mitigates inflammations caused by fire, cures the scab, but especially the decoction being made of its own wood in the same. Being mixed with warm water for a bath for the lower part of the body, it cures occult diseases of women; as also malignant ulcers of the legs.

This spirit therefore deserves some place in the shops, i.e. it is unjustly rejected in the shops, seeing it is easily to be made. In distilling of wormwood and other vegetables, there remains in the bottom of the furnace ashes, which being extracted with warm water yields a salt by decoction, which being again dissolved in its own spirit or vinegar, does by the evaporation of the flegme, being placed in a cold place pass into Crystalline salt, which is of a pleasant taste, not like unto a LIXIVIUM, nor unto other salts that are dissolved in the air. This salt is also more efficacious (being reduced into Crystals by its proper Spirit) than that which is made by the help of sulphur, or Aqua fortis, and oil of Vitriol, and other ways which Chymists, and Apothecaries use.

The Spirit of Paper and Linen Cloth.

Pieces of linen cloth gathered, and got from seamsters being cast into the furnace upon living coals, yield a most acid spirit, which tinges the nails, skin, & hair with a yellow color, restores members destroyed with cold, is good in a gangrene, and erysipelas if linen clothes wet in the same be applied thereto, etc. The same does spirit made of paper, viz. of the pieces thereof.

The Spirit of Silk.

After the same manner is there a spirit made of pieces of silk, which is not so sharp as that which is made of linen and paper, neither does it tinge the skin, but is most excellent in wounds as well old as green, and it makes the Skin beautiful.

The Spirit of Man's Hair, and of Other Animals, as also of Horns.

Out of horns also, and hair is made a spirit, but most fetid, wherefore it is not so useful, although otherwise it may serve for divers arts: being rectified it comes clear and to be of the odor of the spirit of urine. It dissolves common sulphur, and yields a water, that cures the scab in a very short time.

Now for this business, shreds of woolen cloth undyed may serve, being cast in a good quantity into the furnace. Pieces of cloth dipped in this spirit and hung in vineyards, and fields, keep out Deer and Swine from coming in, because they are afraid of the smell of that spirit, as of a huntsman that waits to catch them.

The Spirit of Vinegar, Honey, and Sugar.

He that will distil liquid things, must cast red hot coals into them, as for example into vinegar in the furnace, or if it be honey, or sugar, let them first be dissolved in water, by which means they will be drank up by the coals, which being therewith impregnated, must afterwards at several times be cast into the furnace, and be burnt; and whilst the coals are burning, that which is incombustible comes forth. And by this means you may distil liquid things in a great quantity.

Vinegar which is distilled this way, is of the same nature, as that which is distilled in close vessels.

But honey and sugar that are distilled after this manner, are a little altered, and acquire other virtues: but how they shall be distilled without the loss of their volatile spirit shall be taught in the second part. Also after this manner may all liquid things being drunk up by living coals be distilled.

Of the use of distilled vinegar many things might be said, but because the Books of all the Chymists treat abundantly thereof, I account it needless to repeat what they have writ. Yet this is worth taking notice of, that the sharpest vinegar has a great affinity with some metals, which may be extracted by the help thereof; also dissolved, and reduced into medicaments; yea; many things may be made with the help thereof, as the books of all the Chymists testify.

But there is yet another vinegar, of which there is often mention made in the books of Philosophers, by the help whereof, many wonderful things are performed in the solution of metals, the name whereof the ancients have been silent in; of which I do not here treat, because it cannot be made by this furnace; but I shall treat of it in another part; yet so that I incur not the Curse of the Philosophers.

How Spirits may be made of the Salt of Tartar, Vitriolated Tartar, the Spirit of Salt Tartarizated, and of other such like Fixed Salts.

As many Chymists as there have been, almost all have the opinion that a spirit cannot be drawn out of salt of tartar, and other fixed salts. For experience has taught that by retort little or no spirit can be drawn from thence, as I had often experience of before the invention of this furnace: the reason of which thing was the admixtion of sand, earth, bole, powder of tiles, & etc. for to prevent the flowing of the salt of tartar, being by this means dispersed. But this is done through the ignorance of Authors, who have been ignorant of the properties of salt of tartar. For a stony matter, as sand, flint, bole, & etc. being mixed with salt of tartar, feeling the heat of the fire, and being made red with the same, is joined to it most closely, so as no spirit can be drawn from thence, but become a most hard stone. For sand, and such things that are like to it, have so great an affinity with the salt of tartar, that being once united can scarce ever be separated. Yet it may be made by Art by the addition of pure sand, or flint, because the whole substance of the salt of tartar may be turned into a spirit in the space of one or two hours, as shall be taught in the second part, and it excels all other medicaments in virtue, in curing the stone, and gout. And if by the regiment of art there be left any CAPUT MORTUUM in the distillation, it has, being dissolved in the air, a power to putrefy metals being prepared, and mixed with it, in the space of few hours, to make them become black, and to grow up like trees with their roots, trunks, and boughs, which by how much the longer they are so left, become the better. Of calx of lead being subtilized, and of salt of tartar may be made a SPIRITUS GRADATORINS of wonderful virtues as well in Medicine as Alchemy. There is made of the CAPUT MORTUUM, PER DELIQUIUM a green liquor which does wonderful things; whence it is proved, THAT SATURN IS NOT THE LOWEST OF THE PLANETS; enough to the wise.

And so is the Lac Virgins, and the Philosophical Sanguis Draconis Made.

Sometimes there is found a certain earth, or bole, which has no affinity with tartar, which being mixed with salt of tartar yields a spirit, but very little. But in this furnace, may all fixed things be elevated, because the species not being included in it, but dispersed, being cast upon the fire, are from the fire elevated through the air, and are being refrigerated in the recipients again condensed, which cannot be well done by a close retort.

He that therefore that will make the spirit of the salt of tartar, need do nothing else than cast the calcined tartar into the fire, and it will wholly come over in a spirit; but then there are required glass recipients, because those that are earthen cannot retain it.

And this is the way whereby most fixed salts are distilled into a spirit by the first furnace. In the second furnace (viz. in the furnace of the second part) it may be done better, and easier, where together with the preparation shall be taught the use thereof.

The Spirits, Flowers, and Salts of Minerals and Stones.

By this way spirits may be raised from any mineral or stone, and that without the addition of any other thing: yet so as that the minerals, and stones, as flints, Crystals, talk[7], LAPIS CALAMINARIS, Marcasite, Antimony, being ground with an Iron ladle cast upon the coals, and there will arise together with a certain acid spirit, some salt and flowers, which are to be washed off from the recipients, and filtered, and the flowers will remain in CHARTA BIBULA, or filter for the water together with the spirit, and the salt passes through the filter, all which may be separated, rectified and be kept close by themselves for their proper uses. Now this you must know, that you must choose such minerals which have not been touched by the fire, if you desire to have their spirit.

How Minerals and Metals may be Reduced into Flowers, and of their Virtues.

Hitherto the flowers of metals, and minerals have not been in use, excepting the flowers of

---

[7] The mineral is Talc. The word "talc" is derived from the Persian "tālk." -PNW

Antimony, and sulphur, which are easily sublimed: for Chymists have not dared to attempt the sublimation of other metals, and fixed minerals, being content with the solution of them with Aqua fortis, and corrosive waters, precipitating them with the liquor of salt of tartar, and afterwards edulcorating, and drying them; and being so prepared they have called them their flowers: but by Flowers I understand the same matter which is by the help of fire without the addition of anything sublimed, and turned into a most subtle powder, not to be perceived by the teeth or eyes, which indeed is (in my judgement) to be accounted for the true flowers; when as the flowers which others make are more corporeal, and cannot be so well edulcorated, but retain some saltness in them, as may be perceived by the increase of their weight, and therefore hurtful to the eyes, and other parts.

But our flowers being by the force of the fire sublimed by themselves, are not only without saltness, but are also so subtle that being taken inwardly presently operate, and put forth their powers, viz. according to the pleasure of the Physician, neither is their preparation so costly as the others.

Metals also, and minerals are maturated, and amended in their sublimation, that they may be the more safely taken; but in other preparations they are rather destroyed, and corrupted, as experience witnesses: Now how these kinds of flowers are to be made I shall now teach, and indeed of each metal by itself, whereby the artist in the preparation cannot err, and first thus.

Of Gold and Silver.

Gold and silver can hardly be brought into flowers, because many are of opinion, that nothing comes from them in the fire, especially from Gold, although it should be left there forever: which although it be true, viz. that nothing comes from gold in the fire, although it should remain there a long time, and from silver but a little except it have copper or any other metal mixed, which yet vapors away but by little and little.

Which I say although it be so, yet they being broken and subtilized and scattered upon coals, and so dispersed, may by the force of the fire and help of the air be sublimed, and reduced into flowers.

Now seeing the aforesaid metals are dear, and of a great price, and the furnace with its recipients large, I would not that anyone should cast them in, especially gold, because he cannot recover them all; but I shall to those that desire to make these flowers show another way in the second part, whereby they may make them without the loss of the metal; to which I refer the reader. For this furnace serves for the subliming of metals, and minerals, which are not so precious, the loosing of which, whereof is not so much regarded. And thus, much is said to show that gold, silver, although fixed, may be sublimed. Now other metals may more easily be sublimed, yet one more easily than another, neither need they any other preparation but beating small, before they be cast into the fire.

Flowers of Iron and Copper.

Take of the filings of Iron or Copper, as much as you please, cast them with an Iron ladle upon burning coals, viz. scatteringly, and there will arise from Iron a red vapor, but from Copper a green, and will be sublimed into the sublimatory vessels. As the fire abates it must be renewed with fresh coals, and the casting in of these filings be continued, until you have got a sufficient quantity of flowers, and then you may let all cool. This being done take off the sublimatory vessels, take out the flowers, and keep them, for they are very good if they be mixed with unguents, and emplasters: and being used inwardly cause vomiting; therefore, they are better in Chirurgery, where scarce any thing is to be compared to them. Copper being dissolved in spirit of salt, and precipitated with oil of vitriol, edulcorated, dried, and sublimed, yields flowers, which being in the air resolved into a green balsom, is most useful in wounds, and old putrid ulcers, and is a most precious treasure.

Flowers of Lead and Tin.

You need not reduce these metals into small crumbs, it is sufficient if they be cast in piece by piece, but you must under the grate put an earthen platter glazed, and filled with water, to gather that which flows down melted, which is to be taken out, and cast again into the fire, and this so often until all the metal be turned into flowers, which afterwards are again, the vessels being cold, to be taken out, as has been said of the flowers of MARS and VENUS. And these flowers are most excellent being mixed with plasters and ointments in old and green wounds, for they have a greater power to dry, than metals calcined, as experience can testify.

Of Mercury.

This is easily reduced into flowers, because it is very volatile, but not for the aforesaid reason, because it leaps in the fire, and seeks to descend. And if you desire to have the flowers thereof, mix it first with sulphur that you may pulverize it, and cast into a red-hot crucible set in the furnace, a little quick Mercury, viz. by times with a ladle, presently it will fly out, and some part thereof will be resolved into an acid water, which is to be preferred before the flowers in my judgement; but the rest of the Mercury drops into the receiver. But here are required glass vessels, because the aforesaid water is lost in earthen. And this water without doubt does something in Alchymy.

It is also good being applied outwardly, in the scab, and venereal ulcers.

The Flowers of Zink.

It is a wonderful metal, and is found in the spagyrical anatomy to be mere sulphur, golden, and immature. Being put upon burning coals does suddenly fly away wholly; it is inflamed also, and partly burns like common sulphur, with a flame of another color, viz. golden purple: and yields most gallant white, and light flowers.

The Use.

Being given from 4, 5, 6 grains to 12, they provoke sweat wonderfully, and sometimes vomit, and stools, according to the offending matter. The virtues thereof being externally used are also wonderful, for there are not found better flowers, for they do not only speedily consolidate fresh wounds, but also old, such as always drop water, in which cases they excel all other medicaments. For they are of such dryness, which has joined with it a consolidating virtue, as that they do even things incredible. They may be used divers ways, as to be strewed by themselves, putting over them a styptic plaster, or being brought into an unguent with honey to be put into wounds; which unguents in deep wounds may be boiled to a hardness for the making of small suppositories, which are to be put into the wounds, which must afterwards be covered with some plaster, and preserved from the air. Being applied after this manner they cure fundamentally, being mixed with plaisters also they do wonderful things.

If they be mixed with rose, or rain-water, so as to be united together, and afterwards some of this mixture be sometimes every day dropped into red eyes that water, yielding not to other ophthalmicks, do restore, and heal them.

These flowers being taken up in lint and strewed upon those places of Children that are galled with their urine (those places being first washed with water) heal them quickly. They heal also quickly any excoriation which is contracted by lying long in any sickness, and is very painful, if they be strewed thereon.

These flowers also are more easily dissolved in corrosive waters, than other metals, and minerals, neither does the spirit leave them in the fire, but an insipid phlegm only distils off, leaving a fat and thick oil, as is above said concerning the LAPIS CALAMINARIS, being ordained for the same uses, but more efficacious then that. Which spirit if it be by the violence of fire driven forth, is of so great strength, that it can scarce be kept. And not only spirit of salt, but also Aqua fortis, and Regia may after this manner be exalted, to be able to do wonderful things in the separation of metals; but here is not the place for these things, they shall be spoken of in the fourth part.

But you need not make flowers for this work, because crude Zink does the same, although the flowers do it something better: whence it appears that a metal contracts a higher degree of dryness in sublimation.

Flowers of Antimony.

There is no difficulty to make the flowers of Antimony, for Chymists have a long time made use of them, and because their preparation was tedious, they were not sold at a low rate.

Wherefore there was nobody willing to attempt anything else in them, because they were used only for vomiting; the dose whereof was from 1, 2, 3, 4 grains to 8 and 10 in effects of the stomach and of the head, as also in fevers, plague, morbus gallicus, & etc. Neither is it a wonder if Chymists tried no further in them, for we see that there are found men in these days who persuade themselves that there is nothing which was not found out by the learned ancients, can be found out in these days, and if there were anything to be yet found out it was found out already by them. But this opinion truly is very foolish, as if God gave all things to the ancients, and reserved nothing for them that should come after. Neither indeed do they understand nature in their operations, which works incessantly, and is not wearied in her labours, & etc. But, however, it is manifest that God has revealed things in these times which were hid from them of old, and he will not cease to do the same even to the end of the world.

But to return to our purpose again, which is to show an easier way of making the flowers of Antimony, whereby a greater quantity may be had, as also that they may serve for other uses.

Take of crude Antimony powdered as much as you please, and first make your furnace red hot, then cast in at once a pound of Antimony, or thereabouts, viz. scatteringly upon the coals; and presently it will flow, & being mixed with the coals by the force of the fire will be sublimed through the air into the receivers like a cloud, which will there be coagulated into white flowers. Note, that when the first coals are burnt up, more must be put in to continue the sublimation, and those must be first kindled before they are put in, lest the flowers be by the dust of the coals arising together with them discolored, and contract thence a gray color: but it matters not if you will not use them by themselves to provoke vomiting, because there is no danger thereby, for that color comes only from the smoke of the coals, wherefore you need not be afraid of them. But let him that dislikes this color, first kindle the coals before he puts them into the furnace, and then he shall have white flowers. Also, you must not shut the middle hole through which the coals, and Antimony are cast in, that thereby the fire may burn the more freely: for else the flowers of the superior pots will be yellow and red, because of the sulphur of the Antimony, which is sublimed higher than the regulus. Now you may by this way make a pound of the flowers with 3, 4, or 5 pounds of coals. It is a little that goes away from the Antimony, viz. the combustible sulphur, which is burnt, all the rest going into flowers. You must have a care to provide a sufficient quantity of subliming pots by reason that a large space is required for the sublimation of the flowers.

    The flowers that are prepared after this way, are sold at a lower rate, so that one pound thereof is cheaper, than half an ounce of those that are made after the other manner. Also they are safer, as being made with an open free flame of the fire, for they do not provoke vomit so vehemently; moreover the flowers of the lower pots are not vomiting, but diaphoretical, as if they had been prepared with nitre, for thus they are corrected by the fire: And by this way at one and the same operation divers flowers of divers operations may be made, for the flowers of the lower pots are diaphoretical, of the middle a little vomit—lye, but of the uppermost vehemently vomitive. For by how much the more they have endured the fire, by so much the better are they corrected; from whence the diversity of their power proceeds. Wherefore each of them are to be kept by themselves, and the uppermost for plasters or butter, or oil, and those to be made sweet or corrosive thereby.

The Middle for Purging, and Vomiting, but the Lowermost for Sweat, being

more excellent than BEZOARDICUM MINERALE, or ANTIMONIUM DIAPHORETICUM made with nitre. Truly I do not believe that there is an easier way of making vomiting, and diaphoretical flowers, than ours. Now for the use of them, you must know that those that are vomitive are to be administered to those that are strong, and accustomed to vomit: but to Children, and old Men with discretion, as has been said above of the butter of Antimony: but those that are diaphoretical may be given without danger to Old and Young, to those that are in health, and to the sick; in any affliction that requires sweat; as in the Plague, Morbus Gallicus, Scorbute, Leprosy, Fevers, & etc. The Dose of them is from 3, 6, 9, 12 grains to 24 with proper vehicles to sweat in the bed; for they do expel as well by sweat, as by urine, all evil humours. And because they that are vomitive are in a greater quantity than those that are diaphoretical, and not so necessary as these, and there may be many more doses out of them; it is necessary to show you how those that are vomitive may be turned into diaphoretical; and that may be done three ways; the two former, whereof I have before showed concerning the butter of Antimony made of flowers with spirit of salt, the third is this, viz. put the flowers in a crucible covered (without luting) lest anything fall into it, so set them by themselves in a gentle fire, that they melt not, but be made only darkly glow for the space of some hours; then let them cool, for they are become fixed and diaphoretical. Although they had before contracted some yellowness or ash—color, yet by this means they are made white, fixed, and diaphoretical. Also, these flowers, are used in styptic plasters, because of their dry nature, with which they are endued.

    Also, they are melted into a yellow transparent glass, neither is there taught an easier way of reducing Antimony by itself into a yellow transparent glass, where crude Antimony is first sublimed, and being sublimed Is melted into glass.

    This sublimation serves instead of calcination, by the help whereof 20 pounds are more easily sublimed, than by the help of the other one pound is brought into calx.

Neither is there here any danger of the ascending fumes, because when the Antimony is cast into the fire you may be gone, which is safe, and easy calcination, whereas the common way requires the continual presence of the artist stirring the matter, who also takes out the matter when it is once grown together, and grinds it again; by which means he has much to do, before the matter comes to a whiteness; but by our way, the matter is at the first time made sufficiently white, and more than by that common way of calcination and agitation. I suppose therefore that I have showed to him that will make glass of Antimony, the best, and hitherto unknown way; which being taught, I hope there is no man will hereafter like a fool go that tedious way of the Ancients, but rather follow my steps. For by this way may any Physician, may most easily be able to prepare for himself vomitive and diaphoretical flowers, and also glass of Antimony PER SE.

From those flowers may be made oils both sweet and corrosive, and other medicaments, as has been above said of the spirit of salt, and shall afterwards be spoken in the Second Part.

Let him that will make Flowers of the Regulus, fairer than those which are made of crude Antimony, cast it being powdered into the fire, and in all things, proceed as has been said, and he shall have them, & etc. for they are easily sublimed. Now, how the regulus is to be made after a compendious manner, you shall find in the Fourth Part. The scoriae also are sublimed, so as nothing is lost. But he that will make Flowers that shall be dissolved in the air into a liquor must add some calcined tartar, or some other fixed vegetable salt, and he shall have flowers that will be dissolved in any liquor: but he that will make red flowers as well those that are diaphoretical, as those that are purging, must mix iron, and he shall have flowers like to Cinnabar: Let him that desires green, mix copper, if purple, LAPIS CALAMINARIS.

And thus, out of any mineral may be made flowers whether it be fixed, or volatile; for it is forced to fly on high being cast into the fire. And these may be used diversely in Chirurgery, in plasters and unguents; for they dry, and astringe potently, especially those that are made of LAPIS CALAMINARIS. Neither are they to be slighted that are made of the golden, and silver marcasite. Those that are made of arsenic & auripigmentum, are poisonous, but are useful for Painters. Arsenic & auripigmentum being calcined with nitre, and then sublimed, yield Flowers that are safely to be taken inwardly, expelling all poisons by sweat and stool: For they are corrected two ways, viz. first by the nitre, secondly by the fire in the subliming: they are not therefore to be feared, because that Arsenic was poisonous before the preparation thereof. For by how much the greater poison it was before preparation, so much the greater medicine afterwards.

The Flowers of sulphur are taught in the Second Part, although they may also be made by this furnace, viz. the natures and properties thereof being known by an expert Artist, or otherwise it is burnt.

So also, stones being prepared are brought into Flowers, and many other things, of which we need not say anything, only let him that pleases make trial thereof.

And now I suppose I have made plain, and showed you clearly: how distillation is to be made in this our first furnace; wherefore I will now end. He therefore that understands and knows the fabric of the furnace (which he may understand by the delineation thereof) and the use thereof, will not deny but that I have done a good work, and will not disapprove of my labor.

And this is the best way of distilling, and subliming incombustible things. In the Second Part, you shall find another furnace in which are distilled combustible things, as also most subtle spirits, & etc. The first furnace serves also for other uses, as the separation of metals; of the pure from the impure; for the making of the central salt, and of the HUMIDUM RADICALE of them all. But because it cannot be done after the aforesaid way, by which things, are cast into the fire to get their flowers, and spirits, but after a certain secret Philosophical manner, by the power of a certain secret fire, hitherto concealed by the Philosophers (neither shall I prostrate that secret before all): It is sufficient that I have given a hint of it for further enquiry, and have showed the way to other things.

Finis.

The Second Part of Philosophical Furnaces

Wherein is Described the Nature of the Second Furnace; by the help whereof, all volatile, subtle, and combustible things may be distilled; whether they be Vegetables, Animals, or Minerals, and that after an unknown and very compendious Way; whereby nothing is lost, but even the most subtle spirits may be caught and preserved, which else without the means of this Furnace is impossible to be done by Retorts or other Distilling Instruments.

Of the Structure of the Second Furnace.

The Distilling Vessel must be made of Iron, or good earth, such as can abide in the fire (whereof in the fifth Part of this Book it shall be taught) and you may make it as big or as little as you please, according as your occasion shall require. That of iron is most fit to be used for such spirits, as are not very sharp or corroding, else they would corrode the vessel: but that of earth may be used for such things, as show their activity upon the Iron, and do make it to melt, as sulphur, Antimony and the like; and therefore you ought to have two such vessels, viz. one of iron, and one of earth, to the end that for both sorts of materials (corrosive or not corrosive) you may have proper vessels, and fit furnaces for their distilling, and that they may not be Spoiled by things contrary and hurtful to them. The shape of the vessel is showed by the foregoing figure, viz. the lower part of it somewhat wider than the upper part, and twice as high as wide; at the top having a hollow space between the two edges or brims, whereunto the edge of the lid may close and enter in an inch deep. The lid must have a ring or handle, by which it may be taken off and put on again with a pair of tongs. The lid must have a deep edge answering to the hollow space aforesaid. The lower part must have three knobs or shoulders thereby to rest upon the wall of the furnace; the form whereof is no other, than that of a common distilling furnace with a sand Copple; as the figure of it does show: but if you will not have the furnace, then it needs no knobs or shoulders, if so let the distilling vessel be flat at the bottom, or else have legs, for to stand upon them: Beneath the edge of the vessel there comes forth a spout or pipe of a span in length, and one or two inches wide, and somewhat narrower before than behind, through which the spirits are conveyed into the Receiver.

See the fourth Figure before the first part, wherein the Letter A, represents the Furnace, with the Iron Distilling Vessel fastened into it, whereunto a Receiver is applied.

B. The Distiller, with his left hand taking oft the lid, and with his right-hand casting in his prepared matter.

C. The external form of the distilling vessel.

D. The internal form of the vessel.

E. Another distilling vessel, which is not fastened to a furnace, but only stands upon Coals.

The Way or Manner to Perform the Distillation.

When you intend to distil, then first make a fire in the furnace, that the distilling vessel becomes very hot. But if it be not fastened to the

Furnace, then set it upon a grate, and lay some stones about it, and coals between, and so let it grow hot, and lay melted lead in the space between the two edges or brims, to the end, that the lid, when it is put on, may close exactly, so that no spirit can get through. This done, take a little of the matter you intend to distil, and cast it in, and presently put on the lid, and there will be no other passage left but through the pipe, to which there must be applied and luted a very big receiver. As soon as the species cast in comes to be warm, they let go their spirit, which does come forth into the receiver: and because there was but little of the matter cast in, it has no power to force through the lute, or to break the receiver, but must settle itself. This done, cast in a little more of your matter, cover it and let it go till the spirit be settled: continue this proceeding so long, until you have spirits enough: but take heed, that you cast in no more at once, than the receiver can bear, else it will break. And when your vessel is full, the distillation not being ended, then take off the lid, and with an iron ladle take out the Caput Mortuum; and so began again to cast in, and still but a little at a time, and continue this as long as you please.

    Thus in one day you may distil more in a small vessel, than other ways you could do in a great retort; and you need not fear the least loss of the subtle spirit, nor the breaking of the receiver by the abundance of the spirits: and you may cease or leave off your distilling, and begin it again when you please: also the fire cannot be made too strong, so that it might cause any hurt or damage; but by this way you may make the most subtle spirits, which is impossible to be done by any Retort. But if you distil a subtle spirit by a Retort, as of Tartar, Harts-horn, Sal armoniack, or the like, you cannot do it without prejudice (though there were but half a pound of the matter in it) the subtlest spirits coming forth with force, seek to penetrate through the lute, if that be not good, but if that be good, so that the spirits cannot pass through it, then they break the receiver, because it cannot possibly hold such a quantity of subtle spirits at once. For when they are coming, they come so plentifully, and with such a force, that the receiver cannot contain them, and so of necessity must fly asunder, or must pass through the lutum. All which is not to be feared here, because there is but little cast in at once, which cannot yield such a quantity of spirits, as to force the receiver to break: And when there comes forth no more spirits, and the former is settled, then more of the matter is to be cast in; and this is to be continued so long, until you have spirits enough. Afterward take off the receiver, and put the spirit into such a Glass (as in the fifth part of this book, amongst the Manuals, shall be discovered) wherein it may be kept safely without wasting or evaporating.

In this manner, all things, Vegetable, Animal, or Mineral, may be distilled in this Furnace, and much better, than by means of a Retort: especially such subtle spirits (as by the other way of distilling cannot be saved, but pass through the lutum) are got by this our way; and they are much better than those heavy oils, which commonly are taken for spirits, but are none, being only corrosive waters. For the nature and condition of a spirit is to be volatile, penetrating and subtle, and such are not those spirits of salt, Vitriol, Allome and Nitre, which are used in Apothecary shops, they being but heavy oils, which even in a warm place do not evaporate or exhale.

But a true spirit, fit for Medicinal use, must rise or ascend before the phlegm, and not after; for whatsoever is heavier than flegme is no volatile spirit, but a heavy spirit or (rather called) a sour heavy oil. And it is seen by experience, that the Apothecary's spirit of vitriol will cure no falling sickness, which virtue is ascribed to that spirit, and indeed justly: for the true spirit of vitriol performs that cure out of hand. Likewise, their spirit of Tartar (as they call it) is no spirit, but only a stinking phlegm or vinegar.

The way to make such true spirits, I will now show, because much good may be done by them in all manner of Diseases. And this way of distilling serves only for those which seek after good Medicines: but others which care not whether their medicines be well prepared or not, need not take so much pains as to build such a furnace, and to make their spirits themselves. For at any time they can buy for a small matter, a good quantity of dead and fruitless spirits at the common sellers and Apothecaries.

Hence it is no marvel, that nowadays so little good is done by Chymical medicaments, which of right should far out-strip all the Galenical in goodness and virtue. But alas! It is come to that pass now, that a true Chymist, and honest Son of HERMES, is forced almost to blush, when he hears men talk of Chymical medicines, because they do no such miracles, as are ascribed unto them. Which infamy is occasioned by none more, than by careless Physicians, which though they make use of Chymical medicines, (because they would fain be esteemed to know more than others) yet they do take greater care for their kitchen, than for the welfare of their Patients; and so, buying ill-prepared Medicines of unskillful stillers, and withal using them indiscreetly (whereby they many times do more hurt than good to the sick) they lay such foul aspersions upon the noble Art of Chymistry.

But an industrious and accurate Physician is not ashamed to make his Medicines himself, if it be possible, or at least to have them made by good and well exercised Artists: whereupon he may better rely, and get more credit, than one that knows not whereof, nor how his Medicine which he does administer to his patients is prepared. But such wicked and ignorant men will one day fall short of their answer before the Judgement of the righteous Samaritan.

How to Make the Acid Oil and the Volatile Spirit of Vitriol.

Hitherto I have taught, how to distil in general, and to get the subtle spirits. There remains now to describe what Manuals or Preparations are fitting for every matter in particular; and first;

Of Vitriol.

To distil Vitriol, there needs no other preparation, but only that it be well viewed, and if there be any filth amongst it, that the same be carefully picked out, lest being put together with the Vitriol into the distilling vessel, the spirit be corrupted thereby. But he that will go yet more exactly to work, may dissolve it in fair water, then filtrate it, and then evaporate the water from it till a skin appears at the top, and then set it in a cold place, and let it shoot again into Vitriol; and then you are sure that no impurity is left in it.

Now your vessel being made red hot, with an Iron ladle cast in one or two ounces of your Vitriol at once, put on the lid, and presently the spirits together with the phlegm will come over into the receiver, like unto a white cloud or mist; which being vanished, and the spirits partly settled, carry in more Vitriol, and continue this so long, until your vessel be full: Then uncover your vessel, and with a pair of tongs or an iron ladle take out the Caput Mortuum, and cast more in; and continue this procedure as long as you please, still emptying the vessel when it is filled, and then casting in more matter, and so proceeding until you conceive that you have got spirits enough. Then let the fire go out, and let the furnace cool; take off the receiver, and pour that which is come over into a retort, and set the retort in sand, and by a gentle fire distil the volatile spirit from the heavy oil; having first joined to the retort the receiver, which is to receive the volatile spirit, with a good lutum, such as is able to hold such subtle spirits, the making whereof shall be taught in the fifth part of this Book, amongst the Manuals.

All the volatile spirit being come over, which you may know by the falling of bigger drops, then take off the receiver, and close it very well with wax, that the spirit may not make an escape; then apply another (without luting it) and so receive the phlegm by itself, and there will remain in the retort a black and heavy corrosive oil, which if you please, you may rectify, forcing it over by a strong fire, and then it will be clear; if not, let all cool, then take out your Retort together with the black oil, and pour upon it the volatile spirit, which in the rectifying went over first, put the retort into the sand, and apply a receiver, and give it a very gentle fire, and the volatile spirit will come over alone, leaving its phlegm behind with the oil, which by reason of its dryness does easily keep it. Thus, the spirit being freed from all phlegm, is become as strong as a mere fire, and yet not corrosive. And if this spirit be not rectified from its own oil, it will not remain good, but there does precipitate a red powder after it has stood for some time, and the spirit loses all its virtue, insomuch that it is not to be discerned from ordinary water, which does not happen when it is rectified. The reason of this precipitation is no other than the weakness of the spirit, which is accompanied with too much water, and therefore not strong enough to keep its sulphur, but must let it fall; but after it is rectified by its own oil, it can keep its sulphur well enough, because then it is freed from its superfluous moisture. However, the red powder is not to be thrown away, but ought to be kept carefully; because it is of no less virtue than the spirit itself. And it is nothing else but a Volatile sulphur of Vitriol. It has wonderful virtues, some of which shall be related.

The Use and Dose of the Narcotick Sulphur of Vitriol.

Of this sulphur 1, 2, 3, 4, or more grains (according to the condition of the patient) given at once mitigates all pains, causes quiet sleep; not after the manner of Opium, Henbane, and other like medicines, which by stupefying and benumbing cause sleep, but it performs its operation very gently and safely, without any danger at all, and great Diseases may be cured by the help thereof. PARACELSUS held it in high esteem, as you may see, where he does write of SULPHUR EMBRYONATUM.

Of the Use and Virtue of the Volatile spirit of Vitriol.

This sulphurous Volatile spirit of Vitriol, is of, a very subtle and penetrating quality, and of a wonderful operation; for some drops thereof being taken and sweated upon, does penetrate the whole body, opens all obstructions, consumes those things that are amiss in the body, even as fire. It is an excellent medicine in the falling sickness, in that kind of madness or rage which Is called MANIA, in the Convulsion of the Mother, called SUFFOCATIO MATRICIS, in the Scurvy; in that other kind of madness which is called Melancholia Hypochondriaca and other Diseases proceeding from Obstructions and Corruption of the Blood. It Is also good in the Plague, and all other Fevers: mingled with spirit of wine, and daily used, it does wonders in all external accidents. Also in the Apoplexy, shrinking and other diseases of the Nerves, the distressed limb rubbed therewith, it does penetrate to the very marrow in the bones; it does warm and refresh the cold sinews, grown stiff. In the Cholick, besides the internal use, a little thereof in a clyster applied, is a present help. Externally used in the Gout, by anointing the places therewith, assuages the Pains, and takes away all tumors and inflammations. It does heal scabs, tetters and ring-worms, above all other medicines; it cures new wounds and old sores, as Fistulaes, Cancers, Wolves, and what name so ever else they may have.

It extinguishes all Inflammations, scalding's, the Gangrene, dispatches and consumes the knobs and excrescencies of the skin. In a word, this spirit, which the wise men of old called SULPHUR PHILOSOPHORUM, does act universally in all diseases, and its virtue cannot sufficiently be praised and expressed. And it is much to be admired, that so excellent a Medicine is nowhere to be found.

If it be mingled with Spring water, it does make it pleasantly sourish, and in taste and virtue like unto the natural sour water of wells.

Also by this spirit, many diseases may be cured at home; so that you need not go to baths afar off, for to be rid of them.

Here I could set down a way, how such a spirit may be got in great abundance for the use of bathing, without distillation, whereby miraculous things may be done, but because of the ungratefulness of men, it shall be reserved for another time.

Of the Virtue and Use of the Corrosive Oil of Vitriol.

This oil is not much used in Physick, although it be found almost in every Apothecaries shop, which they use for to give a sourish taste to their syrups and conserves. Mingled with spring water and given in hot diseases, it will extinguish the unnatural thirst, and cool the internal parts of the body. Externally it cleanses all unclean sores, applied with a feather; it separates the bad from the good, and lays a good foundation for the cure.

Also, if it be rectified first, some metals may be dissolved with it and reduced into their Vitriols, especially Mars and Venus; but this is to be done by adding common water thereunto, else it will hardly lay hold on them. The way of doing it is thus.

How to Make the Vitriol of Mars and Venus.

Take of your heavy oil, just as it came over, viz. together with its phlegm (but that the Volatile spirit be drawn off from it first) as much as you please, put it into a glass body together with plates of copper or iron, set it in warm sand, and let it boil until that oil will dissolve no more of the metal, then pour off the liquor, filter it through brown paper, and put it into a low gourd glass, and set it sand, and let the phlegm evaporate until there appears a skin at the top, then let the fire go out, and the glass grow cool; then set it in a cold place, and within some days there, will shoot fair Crystals; if of Iron, greenish; if of Copper, then something blueish; take them out and dry them upon filtering paper, the remaining liquor, which did not shoot into Vitriol, evaporate again in sand, and then let it shoot as before; continue this proceeding, until all the solution (or filtered liquor) be turned to Vitriol. This Vitriol is better and purer than the common; for it yields a better Volatile spirit, and for that reason I did set down the way how to make it. There also may be made a good Vitriol of both these metals by the means of ordinary yellow brimstone; but because the making of it is more tedious, than of this here set down, I think it needless to describe its preparation in this place.

The Way to Make a Fair Blue Vitriol out of LUNA (that is, Silver.)

Dissolve the shavings or filings of silver with rectified oil of Vitriol, adding water thereunto, but not so much as to Iron or Copper. Or else, which is better, dissolve calcined silver, which has been precipitated out of Aqua fortis either with Copper or salt water; the solution being ended pour it off and filter it, and drop into it of spirit of urine or Sal armoniack, as long as it does hiss, and almost all the silver will precipitate again out of the oil, and so there will fall a white powder to the bottom. This precipitated silver together with the liquor pour into a phial-glass, set it to boil in sand for twenty-four hours, and the liquor will dissolve again almost all the precipitated silver—calx and become blue thereby. Then pour off the solution (or liquor) and filter it through brown paper, and abstract the moisture till a skin rises at the top; then in a cold place let it shoot to Vitriol. With the remaining liquor proceed further, as above in the preparation of the Vitriol of Iron and Copper has been taught.

By this method, you will get an excellent Vitriol out of silver, which from 4, 5, 6, to 10 grains used only of itself, will be a good purge, especially in diseases of the brain.

If you have a good quantity of it, that you may distill a spirit thereof, you will get not only an acid (or sour) but also a volatile spirit, which in the infirmities of the brain is most excellent; that which in the distilling remains behind, may be reduced again into a body, so that you lose nothing of the silver, save only that which is turned into spirit.

Moreover, the acid (or sour) oil of common Vitriol, does precipitate all metals and stones of beasts or fishes; also pearls and corals, they being first dissolved in spirit of salt or of Nitre, and make fair light powders of them (which by the Apothecaries are called Magisteries) much fairer than by precipitation with salt of Tartar is done, especially of corals and pearls, such a fair glistering and delicate powder is made; and likewise also of mother of pearl, and other shells of snails, that it gives as fair a gloss to them, as the fairest oriental pearls have; which way has not been made common hitherto, but being known only to few, has been kept very secret by them, as a singular Art. Such Magisteries commonly were precipitated out of vinegar only by salt of Tartar, which for lightness, whiteness and fair gloss are not comparable at all to ours. But if instead of the oil of Vitriol you take oil of sulphur, then these powders will be fairer than when they are done by the oil of Vitriol, in so much, that they may be used for painting for a black skin.

Having made mention of Magisteries, I cannot forbear to discover the great abuse and error, which is committed in the preparing of them.

PARACELSUS in his **Archidoxes** teaches to make Magisteries, which he calls extracted Magisteries: but some of his disciples teach to make precipitated Magisteries which are different from the former. PARACELSUS is clean of another opinion in the preparing of his Magisteries, then others in the making of theirs: doubtless PARACELSUS his Magisteries were good cordial living medicines, whereas the other are but dead carcasses, and although they be never so fair, white and glistering, yet in effect they prove but a gross earthly substance, destitute of virtue.

    I do not deny, but that good medicines may be extracted out of pearls and corals, for I myself also do describe the preparations of some of them; but not at all after such a way as theirs is. For what good or exalting can be expected by such a preparation, where a stony matter is dissolved in corrosive waters, and then precipitated into stone again? Can its virtue be increased thereby? Surely not, but rather it is diminished, and made much the worse thereby. For it is well known, that the corrosive spirits (no less than fire) do burn some certain things; for not all things are made better by fire or corrosives, but most of them are absolutely spoiled by them. Some perchance will say, that such preparations of Magisteries are only to be reduced into a finer powder, that so much the sooner they may perform their operation. To which I answer, that pearls, corals, and other things of the like nature, if they be once dissolved by corrosive waters, and then precipitated and edulcorated, never or hardly can be dissolved again by acid spirits. Whence it is evident that by such preparations they are not opened or made better, but rather closed or made worse. And we see also by daily experience that those Magisteries do not those effects, which are ascribed unto them. By which it appears clearly, that to the Archeus of the stomach they are much less grateful than the crude unprepared corals and pearls; whose tender essence being not burnt up by corrosives, do oftentimes produce good effects. For our Ancestors have ascribed unto corals and pearls, that they purify the impure and corrupt blood in the whole body, that they expel Melancholy and sadness, comforting the heart of man, and making it merry, which also they effectually perform: whereas the Magisteries do not. And this is the reason, why unprepared corals, pearls and stones of fishes have more effect, than the burnt Magisteries. For it is manifest and well known, that the above said diseases for the most part do proceed from obstructions of the spleen, which obstructions are nothing else, but a tartarous juice or a sour flegme which has possessed and filled up the entrails, and coagulated itself within them. By which obstruction not only headache, giddiness, panting of the heart, trembling of the limbs, a spontaneous lassitude, vomits, unnatural hunger; also, loathing of victuals; then hot flushing fits, and many more strange symptoms are caused; but also, a most hurtful rottenness and corruption is introduced into the whole mass of blood, from whence the leprosy, scurvy, and other loathsome or abominable scabs do spring.

Of which evil, the only cause (as has been said) is a crude acid Tartar, from which so many great diseases do rise.

This to be so may easily be proved; for it is notorious, that melancholic folks, hypocondriaques, and others do often cast up a great quantity of acid humour, which is so sharp that no vinegar is comparable to it, and does set their teeth on such an edge, as if they had eaten unripe fruit.

What remedy now? Take away the cause and the disease is taken away. If you could take away the peccant matter by purgings, it would be well, but it remains obstinate and will not yield to them. By vomit it may be diminished in some measure. But because that not everyone can abide vomiting, it is therefore no wisdom to turn evil into worse. Shall then this tartar be killed and destroyed by contraries, which indeed in some sort may be effected; as when you use vegetables or animals, whose virtue consists in a volatile salt: such are all species or sorts of cresses, Mustard-seed, horseradish, scurvy grass, also the spirit of Tartar, or Harts-horn, and of urine and the like, which by reason of their penetrating faculty pass through all the body, finding out the Tartar thereof, destroying the same, as being contrary unto it; and in this combat two contrary natures are kindled, a great burning heat, whereby the whole body is thoroughly heated and brought to sweating; and whenever by these contraries a sweating is caused, there is always mortified some of this hurtful Tartar. But because of that acid humor, but a little at a time can be mortified and edulcorated by contrary volatile spirits, and that therefore it would be required to use them often, for to kill and expel all the Tartar; and because also (as has been mentioned before) a strong sweat always is caused by every such operation, whereby the natural spirits are much weakened, so that the patient would not be able to hold out long, thereby, but by taking away of one evil, another and greater one would be occasioned.

And therefore, such things must be offered to that hungry acid humour, by which the corrosive nature thereof, may be mortified and grow sweet, with that proviso nevertheless, that those things be such as are not contrary or hurtful to the nature of man, but grateful and friendly, as are corals, pearls and crab's eyes, & etc.

For amongst all stones none are more easily to be dissolved than Pearls, Corals, Crabs eyes, and other stones of fishes.[8]

---

[8] Such stones are made up of calcium carbonate, $CaCO_3$. -PNW

But the truth of this, viz. that every corrosive is killed by feeding upon pearls and corals, and thereby can be made sweet; and besides, how a sour coagulated Tartar, by the help of corals or pearls may be reduced to a sweet liquor (a pleasant and acceptable medicine to the nature of man) which never can be coagulated again by any means, shall be afterwards proved and taught when I shall come to treat of Tartar.

Now in tartareous coagulations and obstructions of the internals proceeding from the predominance of an acid humor, there is no better remedy than to give the patient every morning fasting from 1/24 ounce to 1/8 ounce (more or less, according to the condition of the patient) of red corals and pearls[9] made into powder, and let him fast two or three hours upon it, and so continue daily until you see amendment. By this means the hurtful acid humor is mortified, and dulcified by the corals and pearls, so that afterwards it may be overcome by nature, whereby the obstructions are removed, and the body freed from the disease.

This my opinion of the abuse of Magisteries and the good use of corals I could not conceal, although I do know for certain, that it will take but with few, in regard that it will seem very strange to most. However, happily there may be some yet, that will not be unwilling to search into the truth, and to consider further of it, and at last will find this not to be so strange, as it seemed to them at first; but he that cannot believe or comprehend it, may keep to his Magisteries.

---

[9] Calcium carbonate is widely used medicinally as an inexpensive dietary calcium supplement for gastric antacid. *Medline Plus*, National Institutes of Health, Oct. 1, 2005.

And if it seems so strange unto any, that corals or pearls made into powder shall be concocted in the stomach, and so put forth their virtue, what will you say then, if I do prove, that even whole pearls, crabs-eyes, and corals being swallowed, are totally consumed by the Melancholy humor, so that nothing comes forth again among the excrements? And which is more, even the like may be said of hard and Compact metals, as Iron, and Speaucer or Zink: But this must be understood only of those that are of a Melancholic constitution but not so in others, viz. those that are of a sanguine, and those that are of a phlegmatic constitution, to whom such like things are seldom prescribed. For I have seen many times, that against obstructions, to strong bodies there has been given at once from 1/24 to 1/8 ounce of the shavings or filings of Iron, and they found much good by it, yea more help then by other costly medicines of the Apothecaries, whereof they had used many before, but to no purpose, by reason whereof their excrements came from them black, just as it uses to fall out with those that make use of medicinal sour waters, which run through iron mines, and thereby borrow a spiritual mineral virtue.

Now if those filings of iron had not been consumed in the stomach, how come it that the excrements are turned black? So, then it is sufficiently proved, that even a hard-unprepared metal can be consumed in the stomach: and if so, why not as well soft pearls and corals?

Which is also to be seen by children, that are troubled with worms, if there be given unto them 4, 6, 8 to 16 grains of the finest filings of steel or iron, that all the worms in the body are killed thereby, their stomach and guts scoured very clean, and their stools also turned black. But this must be observed by children, when the worms are killed, and yet remain in the guts (because that the iron in a small quantity is not strong enough for to expel them, but only make the body soluble) that a purge must be used after, for to carry them out; for else if they do remain there, others will grow out of their substance. But to those that are more in years, you may give the Dose so much the stronger, as from 1/24 to 1/8 ounce that the worms also may be carried out, they being better able to endure it than little children, and although sometimes a vomit does come, yet it does no hurt, but they will be but so much the healthier afterward.

And thus, Iron may be used, not only against worms, but also against all stomach-agues, head-ache, and obstructions of the whole body, without any danger and very successfully, as a grateful or very acceptable medicine to Nature; for after a powerful magnetical way it does attract all the ill humours in the body; and carries them forth along with it. Of whose wonderful virtue and nature, there is spoken more at large in my Treatise of the Sympathy and Antipathy of things. Which some Physicians perceiving and supposing by Art to make it better, they spoiled it, and made it void of all virtue: for they taking a piece of steel, made it red-hot, and held it against a piece of common Sulphur, whereby the steel grew subtle, so they did let it drop into a vessel filled with water; then they took it out, and dried it, and made it into powder, and used it against obstructions, but to no effect almost; for the Iron was so altered by the sulphur, and reduced to an insoluble substance (which ought not to have been so) that it could perform no considerable operation: But if they had made the steel more soluble (whereas they made it more insoluble) than it was of itself before, then they had done a good work: for he that knows sulphur, does know well enough, that by no AQUA FORTIS or AQUA REGIS it can be dissolved; and how could it then be consumed by an animal humor?

Hitherto it has been proved sufficiently, that in some men, especially in those that are of a Melancholic constitution there is an acid humor, which can sufficiently dissolve all easily soluble metals and stones: and that therefore it is needless to torture, and dissolve pearls, corals and the like with corrosive waters before they be administered to patients: but that the Archeus of the stomach is strong enough by the help of the said humours to consume those easily soluble things, and to accept of that which serves his turn, and to reject the rest.

But it is not my intent here, that this should be understood of all metals and stones, for I know well, that other metals and stones (some excepted) before they are duly prepared, are not fit for Physick, but must be fitted first, before they be administered or given unto patients.

For this relation, I made only to show how sometimes good things (though with intent to make them better) are made worse, and spoiled by those that do not make an exact search into nature and her power.

I hope this my admonition will not be taken ill, because my aim was not vain—glory, but only the good of my neighbor.

Now let us again return to Vitriol.

Of the Sweet Oil of Vitriol.

The Ancients make mention of a sweet and green oil of Vitriol, which does cure the falling sickness, kills worms, and has other good qualities and virtues besides: and that the Oil is to be distilled PER DESCENSION. To attain unto this oil the latter Physicians took great pains, but all in vain: because they did not understand at all the Ancients about the preparing of this oil, but thought to get it by the force of fire, and so using violent distillations, they got no sweet oil, but such as was very sour and corrosive, which in taste, efficacy and virtue, was not comparable at all to the former.

However, they ascribed unto it (though falsely) the same virtues, which the ancients (according to truth) did unto theirs. But daily experience shows, that the oil of vitriol as it is found ordinarily, cures no falling sickness, nor kills worms, whereas this Philosophical does it very quickly. Whence it appears, that the other is nothing like unto the true medicinal oil of vitriol, neither is it to be compared to it.

I must confess indeed, that PER DESCENSION out of common vitriol, by the force of the fire, there may be got a greenish oil, which yet is not better than the other, because it proves as sharp in taste, and of as corroding a quality, as if it had been distilled through a Retort.

Those that found out this oil, as PARACELSUS, BASILIUS, and some few others, did always highly esteem it, and counted it one of the four main pillars of Physick. And PARACELSUS says expressly in his writings, that its viridity or greenness must not be taken away or marred (which indeed a very little heat can do) by the fire, for (says he) if it be deprived of its greenness, it is deprived also of its efficacy and pleasant essence. Whence it may be perceived sufficiently, that this sweet green oil is not to be made by the force of the fire as hitherto by many has been attempted, but in vain.

And it is very probable, that the ancients, which did so highly praise the oil of vitriol, happily knew nothing of this way of distilling, which is used by us now: for they only simply followed Nature, and had not so many subtle and curious inventions and ways of distilling.

But it is certain that such a sweet and green oil cannot be made of vitriol by the force of the fire, but rather must be done by purification, after a singular way; for the Ancients, many times understood purification for distillation: as it is evident, when they say, distill through a Filter, or through filtering paper: which by us is not accounted for distillation, but by them it was.

However, this is true and very sure, that a great Treasure of health (or for the health of man) lies hidden in Vitriol: yet not in the common, as it is sold everywhere, and which has endured the heat of the fire already; but in the Ore as it is found in the earth, or its mine. For as soon as it comes to the day light, it may be deprived by the heat of the Sun of its subtle and penetrating spirit, and so made void of virtue; which spirit, if by Art it be got from thence, smells sweeter then musk and amber, which is much to be admired, that in such a despicable mineral and gross substance (as it is deemed to be by the ignorant) such a royal medicine is to be found.

Now this preparation does not belong to this place, because we treat here only of spirits, which by the force of fire are driven over. Likewise also, there does not belong hither the preparation of the green oil, because it is made without the help of fire. But in regard, that mention has been made of it here, I will (though I kept it always very secret) publish it for the benefit of poor patients, hoping that it will do much good to many a sick man.

For if it be well prepared, it does not only cure perfectly every Epilepsy or Convulsion in young and old; and likewise readily and without fail kills all worms within and without the body, as the Ancients with truth ascribed unto it; but also many Chronical diseases and such as are held incurable, may be happily overcome and expelled thereby, as the plague, pleurisy, all sorts of fevers and agues, whatever they be called, head-ache, collick, rising of the mother; also all obstructions in the body, especially of the spleen and liver, from whence MELANCHOLIA HYPOCHONDRIACA, the scurvy, and many other intolerable diseases do arise: Also the blood in the whole body is by the means thereof amended and renewed, so that the Pox, Leprosie, and other diseases proceeding from the infection of the blood are easily cured thereby: Also it heals safely and admirably all open sores and stinking ulcers turned to fistula's in the whole body, and from what cause so ever they did proceed, if they be anointed therewith, and the same also be inwardly used besides.

Such and other diseases more (which it is needless here to relate) may be cured successfully with this sweet oil; especially, if without the loss of its sweetness it be brought to a red color; for then it will do more than a man dare write of it, and it may stand very well for a PANACIA in all diseases.

The Preparation of the Sweet Oil of Vitriol.

Commonly in all fat soils or clay grounds, especially in the white, there is found a kind of stones, round or oval in form, and in bigness like unto a pigeons or hens—egg, and smaller also, viz. as the joint of one's finger, on the outside black, and therefore not esteemed when it is found, but cast away as a contemptible stone. Which if it be cleansed from the earth, and beaten to pieces, looks within of a fair yellow and in streaks, like a gold Marcasite, or a rich gold Ore, but there is no other taste to be perceived in it, then in another ordinary stone; and although it be made into powder, and boiled a long time in water, yet it does not alter at all, nor is there in the water, any other taste or color, than that which it had first (when it was poured upon the stone) to be perceived. Now this stone is nothing else, but the best and purest Mineral (or Ore) of Vitriol, or a seed of Metals; for Nature has framed it round, like unto a vegetable seed, and sowed it into the earth, out of which there may be made an excellent medicine, as follows.

Take this Ore or Mineral beaten into pieces, and for some space of time, lay or expose it to the cool air, and within twenty or thirty days it will magnetically attract a certain saltish moisture out of the air, and grow heavy by it, and at last it falls asunder to a black powder, which must remain further lying there still, until it grows whitish, and that it does taste sweet upon the tongue like vitriol. Afterward put it in a glass—vessel, and pour on so much fair rain water, as that it cover it one or two inches; stir it about several times a day, and after a few days the water will be colored green, which you must pour off, and pour on more fair water, and proceed as before, stirring it often until that also come to be green: this must be repeated so often, until no water more will be colored by standing upon it. Then let all the green waters which you poured off, run through filtering paper, for to purify them; and then in a glass—body cut off short let them evaporate till a skin appears at the top: then set it in a cold place, and there will shoot little green stones, which are nothing else but a pure vitriol: the remaining green water evaporate again, and let it shoot as before: and this evaporating and Crystallizing must be continued until no vitriol more will shoot, but in warm and cold places there remains still a deep green pleasant sweet liquor or juice: which is the true sweet and green oil of Vitriol, and has all the virtues above related.

But now this green oil further without fire may at last (after the preparing of many fair colors between) be reduced to a blood red, sweet and pleasant oil, which goes far beyond the green both in pleasantness and virtue, and is in comparison to it like a ripe grape to an unripe: Hereof happily shall be spoken at another time, because occasion and time will not permit me now to proceed further in it. And therefore, the Philo—Chymical Reader is desired for the present to be contented with the green oil, to prepare it carefully, and to use it with discretion; and doubtless he will get more credit by it, and do more wonderful things then hitherto has been done by the heavy corrosive oil.

The Use and Dose of the Sweet Oil of Vitriol.

Of this green oil, there may be taken from 1, 2, 4, 8, 10 or 12 drops at once, according to the condition of the patient and the disease, in fit Vehicles, in Wine or Beer, in the morning fasting, as other medicines are usually taken. Also, the Dose may be increased or lessened, and as often reiterated as the disease shall require.

This Oil expels all ill humours, not only by stool and vomits, but also by urine and sweating, according as it does meet with superfluities; and this very safely, and without any danger at all; whereby many diseases radically or perfectly may be cured.

Let no man wonder that I ascribe such great virtues unto this oil, it coming from such a despicable stone, and its preparation requiring no great Art or pains, as those intricate deceitful processes do, that are everywhere extant in books quite filled up with them. And it is no marvel, that men are in love with such false and costly processes; for the most of them do not believe, that any good is to be found in things that are not in esteem; but only make great account of dear things, farfetched, and requiring much time and pain to be prepared.

Such men do not believe the word of God, testifying, THAT GOD IS NO RESPECTER OF PERSONS, but that all men that fear and love him, are accepted of him. If this be true (which no good Christian will doubt) then we must believe also, that God created Physick or the matter of Physick as well for the poor as for the rich. Now if it be also for the poor, then certainly such will be the condition thereof, that it may be obtained by them, and easily prepared for use. So, we see that Almighty God causes not only in great men's grounds to come forth good Vegetables, Animals and Minerals, for the curing of the infirmities of mankind, but that the same also are found everywhere else. Whereby we perceive, that it is also the will of God, that they shall be known by all men, and that he alone, as the Maker of all good, may be praised and magnified by all men for the same.

I doubt not but there will be found self-conceited scoffers, that will despise this so little regarded subject, as if no good thing could be made of it, because they could find nothing in it themselves. But be it known to them, that neither to me nor them all things have been discovered, but that yet many wonderful works of Nature are hidden to us: and besides that, I am not the first that writ of Vitriol and its medicine. For the Ancients, our dear Ancestors, had always Vitriol in very great esteem, as the following Verse proves.

Visitabis Interiora Terra, Rectificando Invenies Occultum Lapidem, Veram Medicinam.

Thereby they would give us to understand, that a true medicine is to be found in it. And the same also was known to the latter Philosophers: for BASILIUS and PARACELSUS have always highly commended it, as in their writings is to be found.

It is to be admired, that this Ore or Metallical seed, which may justly be called the gold of Physicians (in regard that so good a medicine can be made of it) is not changed or altered in the earth, like other things that grow in it, but keeps always the same form and shape, until it comes to the air, which is its earth or ground, wherein it putrefies and grows. For first it swells and grows like as a vegetable seed does in the earth: and so, takes its increase and grows out of the air, just as a seed of an herb in the earth; and the air is not only its Matrix, wherein it grows and does increase like a vegetable, but it is also its Sun which makes it ripe. For within four weeks at the furthest it putrefies and grows black: and about a fortnight after it grows white, and then green; and thus, far it has been described here: But if you proceed further Philosopher-like therewith, there will come forth to light at the last the fairest red, and most pleasant Medicine, for which God be praised for ever and ever Amen.

Of the Sulphureous Volatile and Acid Spirit of Common Salt, and of Allome.

The same way, which above has been taught for the making of the volatile spirit of vitriol, must be likewise used in the making of the volatile spirits of common salt and allome.

The Manner of Preparing.

Allome is to be cast in as it is of itself, without mixing of it, but salt must be mixed with bole, or some other earth, to keep it from melting: with the spirit volatile, there goes also along an acid spirit, whose virtue is described in the first part. The Oil of Allome has almost the like operation with the oil of vitriol. Also, the spirit volatile of both these, is of the same nature and condition with that which is made of vitriol: but common salt, and allome, do not yield so much as vitriol; unless both, viz. salt and allome be mixed together, and so a spirit distilled of them.

Of the Sulphureous Volatile Spirit of Minerals and Metals, and of their Preparation.

Such a penetrative sulphureous spirit may be made also of Minerals and Metals, which in virtue goes beyond the spirit of vitriol, that of common salt, and that of allome, viz. after the following manner.

The Preparation of the Volatile Spirits of Metals.

Dissolve either Iron or Copper, or Lead or Tin with the acid spirit of vitriol, or of common salt: abstract or draw off the phlegm; then drive the acid spirit again from the Metal, and it will carry along a volatile spirit, which by rectifying must be separated from the corrosive spirit. And such Metallical spirits are more effectual than those that are made of the salts.

The Preparation of the Volatile Spirit of Minerals.

Take of Antimony made into fine powder, or of golden Marcasite, or of some other sulphureous Mineral, which you please, two parts, mix therewith one part of good purified Salt nitre, and cast in of that mixture a small quantity, and then another, and so forth after the manner above described; and there will come over a spirit which is not inferior to the former in efficacy and virtue; but it must also be well rectified.

Another Way.

Cement what laminated or granulated Metal you please, (except gold) with half as much in weight of common sulphur, closed up in a strong melting pot or crucible, such as does not let the sulphur go through, for the space of half an hour, until that the sulphur has penetrated and broken the plates of Metal: Then beat them into powder, mix them with the like quantity in weight of common salt, and so distil it after the way above mentioned, and you will get a volatile spirit of great virtue: and every such spirit is to be used for such special part or member of the Body, as the Metal is proper for, out of which the spirit is made. So, silver for the brain; Tin for the lungs, Lead for the spleen, and so forth.

The Spirit of Zink.

Of Zink, there is distilled both a volatile and an acid spirit, good for the heart; whether it be made by the help of the spirit of vitriol, or of salt, or of allome: or else by the means of Sulphur; for Zink is of the nature of gold.

The Volatile Spirit of the Dross of Regulus Martis.

The black scoria of the REGULUS MARTIS, being first fallen asunder in the air, yields likewise a very strong sulphureous volatile spirit, not so much unlike in virtue unto the former.
The like Sulphureous volatile spirits may be made also of other minerals, which for brevities sake we omit, as also in regard, that they are almost the same in virtue.

How to Make a White Acid, and a Red Volatile Spirit out of Salt Nitre.

Take two parts of Allome, and one part of salt nitre, make them both into powder, mix them well together, and cast into the still a little and a little thereof, as above in the making of other spirits has been taught, and there comes over an acid spirit together with the volatile spirit; and so many pounds as there is of the materials, which are to be cast in, so many pounds of water must be put into the receiver, to the end that the volatile spirits may so much the better be caught and saved. And when the distillation is performed, the two spirits may be separated by the means of a gentle rectification made in BALNEO; and you must take good heed, that you get the volatile spirit pure by changing the receiver in good time, so that no flegme be mixed with the red spirit, whereby it will be weakened and turn white. The mark whereby you may perceive, whether the spirit or the flegme does go forth is this: when the volatile spirit goes, then the receiver looks of a deep red: and afterward when the flegme does come, the receiver looks white again: and lastly, when the heavy acid spirit goes, then the receiver to be red again, but not so as it was, when the first volatile spirit came over.

This spirit may also be made and distilled after another way, viz. mixing the salt nitre with twice as much bole or brick dust, and so framed into little balls to prevent melting; but no way is so good as the first; especially when you will have the red volatile spirit.

Of the Use of the Red Volatile Spirit.

This volatile spirit, which (being quite freed from flegme) remains always red, and does look like blood, in all occasions may be accounted like in virtue unto the former sulphureous spirits, especially in extinguishing of inflammations and Gangrenes it is a great treasure, clothes being dipped in it, and laid upon the grieved place. Also, it goes almost beyond all other medicines in the Erysipelas and colick: and if there be any congealed blood in the body (which came by a fall or blow) this spirit outwardly applied with such waters as are proper for the grief, and also taken inwardly, does dissolve and expel it: and being mingled with the volatile spirit of urine it does yield a wonderful kind of salt, as hereafter shall be taught.

The Use of the White Acid Spirit of Salt Nitre.

The heavy and corrosive spirit of salt nitre is not much used in Physick, though it be found almost in all Apothecaries shops, and there is kept for such use, as above has been mentioned of the spirit of vitriol, viz. to make their conserves, and cooling-drinks taste sourish.

Also it is used by some in the colick, but it is too great a corrosive, and too gross to be used for that purpose; and although its corrosiveness may be mitigated in some measure, by adding of water thereto, yet in goodness and virtue it is not comparable at all to the volatile spirit, but is as far different from it, as black from white, and therefore the other is fittest to be used in Physick; but this in dealing with metals and minerals, for to reduce them into vitriols, calxes, flores, and crocus.

Aqua Regis.

If you dissolve common salt (which has been decrepitated first) in this acid spirit of salt nitre, & rectify it by a glass retort in sand) by a good strong fire, it will be so strong, that it is able to dissolve gold, and all other metals and minerals, except silver and sulphur; and several metals may by the means thereof be separated much better than by that Aqua Regis which has been made by adding of Salt Armoniack. But if you rectify it with LAPIS CALAMINARIS or Zink, it will be stronger yet, to be able to dissolve all metals and Minerals (silver and sulphur excepted) whereby in the handling of Metals, much more may be effected, than with common spirit of salt nitre or Aqua fortis, as hereafter shall be taught: and first in the preparing of gold.

The Preparation of Aurum Fulminans, or Aurum Terrestrams.

Take of fine granulated or laminated gold (whither it be refined by Antimony or AQUA FORTIS) as much as you please: put it in a little Glass body, and pour four or five times as much of AQUA REGIS upon it, set it stopped with a Paper in a gourd in warm sand; and the AQUA REGIS within the space of one or two hours will dissolve the gold quite into a yellow water: but if it have not done so, it is a sign that either the water was not strong enough, or that there was too little of it for to dissolve it. Then pour the solution from the gold, which is not dissolved yet, into another glass, and pour more of fresh AQUA REGIA upon the gold: set it again to dissolve in warm sand or ashes, and the remaining gold will likewise be dissolved by it, and then there will remain no more, but a little white calx, which is nothing else but silver, which could not be dissolved by the AQUA REGIA (for the AQUA REGIA, whether it be made after the common way with salt Armoniack, or else with common salt, does not dissolve silver) so in like manner common AQUA FORTIS, or spirit of salt nitre dissolves no gold; but all other metals are dissolved as well by strong AQUA FORTIS as by AQUA REGIS. And therefore you must be careful to take such gold as is not mixed with Copper, else your work would be spoiled: for if there were any Copper mixed with it, then that likewise would be dissolved and precipitated together with the gold; and it would be a hindrance to the kindling or fulminating thereof: but if you can get no gold, that is without Copper, then take Ducats or Rose-nobles, which ought to have no addition of Copper, but only of a little Silver, which does not hurt, because that it cannot be dissolved by the AQUA REGIA, but remains in the bottom in a white powder. Make those Ducats or Rose-nobles red hot, and afterward bend them and make them up in Rolls, and throw them into the AQUA REGIA for to dissolve. All the gold being turned into a yellow water, and poured off, pour into it by drops of pure oil made of the Salt of Tartar, PER DELIQUIUM, and the gold will be precipitated by the contrary liquor of Salt of Tartar into a brown yellow powder, and the solution will be clear. But you must take heed, to pour no more oil of Tartar into it than is needful for the precipitation of the gold; else part of the precipitated gold would be dissolved again, and so cause your loss. The gold being well precipitated, pour off the clear water from the gold calx by inclination, and pour upon it warm rain or other sweet water, stir it together with a clean stick of wood, and set it in a warm place, until the gold is settled, so that the water stands clear upon it again; then pour it off, and pour on other fresh water, and let it extract the saltness out of the gold calx; and this pouring off, and then pouring on of fresh water again, must be reiterated so often, until no sharpness or saltness more be perceived in the water that has been poured off. Then set the edulcorated gold into the Sun or another warm place to dry. But you must take heed that it have no greater heat than the heat of the Sun is in MAY or JUNE, else it would kindle or take fire, and (especially if there be much of it) give such a thunder-clap, that the hearing of those that stand by, would be much endangered thereby, and therefore I advise you to beware, and cautious in the handling of it, lest you run the hazard both of your gold and of your health by your oversight.

There is also another way for to edulcorate your precipitated gold, viz. thus. Take it together with the salt liquor, and pour it into a funnel lined with brown Paper laid double, and so let the water run through into a glass vessel, whereupon the funnel does rest, and pour on other warm water, and let it run through likewise; do this again and again, until that the water come from it as sweet as it was poured on. Then take the Paper with the edulcorated calx, out of the funnel, lay it, together with the paper, upon other brown paper lying severally double together, and the dry paper will attract all the moistness out of the gold calx, so that the gold will be dried faster. Which being dry, take it out of the filtering paper, and put it into another that is clean, and so lay it aside, and keep it for use. The salt water that came through by filtering, may be evaporated in a little glass body (standing in sand) to the dryness of the salt, which is to be kept from the air: for it is likewise useful in Physick; because some virtue of the nature of gold is yet hidden in it: though one would not think it, in regard that it is so fair, bright and clear. Which for all that may be observed by this, that when you melt it in a clean covered crucible or pot, and pour it afterward into a clean Copper mortar or basin (being first made warm) you get a purple-colored salt, whereof 6, 9, 12, to 24 grains given inwardly, does cleanse and purge the stomach and bowels, and especially it is used in fevers and other diseases of the stomach. But in the crucible, out of which the salt has been poured, you will find an earthy substance, which has separated itself from the salt, and looks yellowish; this being taken out and melted in a little crucible by a strong fire, turns to a yellow glass, which is impregnated with the Tincture of Gold, and does yield a grain of Silver in every regard like unto common cupellated silver, wherein no gold is found, which is to be admired: because that all Chymists are of opinion, that no AQUA REGIA can dissolve silver which is true. The question therefore is, from whence or how this silver came into the salt, since no AQUA REGIA does dissolve silver? Whereupon some perchance may answer, that it must have been in the oil of Tartar, in regard that many do believe, that the salts likewise may be turned into metals, which I do not gainsay, but only deny that it could have been done here; for if that silver could have been existing in the AQUA REGIA, or salt of Tartar (whereas AQUA

REGIA cannot bear any) it would have been precipitated together with the gold. But that it was no common silver, but gold which turned to silver after it was deprived of its Tincture, I shall briefly endeavor to prove. For that the salt waters (of AQUA REGIA and salt of Tartar) out of which the gold has been precipitated, is of that nature, before it be coagulated to salt, though it be quite clear and white, that if you put a feather in it, it will be dyed purple within few days, which purple color comes from the gold, and not from silver, in regard that silver does dye red or black: and hence appears, that the salt water has retained something of gold.

Now somebody may ask, if that the said salt water has retained some gold, how is it then, that in the melting, no gold comes forth, but only silver? To which I answer, that some salts are of that nature, that in the melting, they take from gold its color and soul; whereof if the gold be truly deprived, it is then no more gold, nor can be such; neither is it silver, but remains only a volatile black body, good for nothing, which also proves much more unfixed than common Lead, not able to endure any force of fire, much less the cupel: But like MERCURY or ARSENIC vanishes (or flies away) by a small heat. Hence it may be gathered, that the fixedness (or fixity) of gold does consist in its soul or Tincture, and not in its body, and therefore it is credible, that gold may be anatomized, its best or purer part separated from the grosser (or courser) and so that a Tingent medicine (or Tincture) may be made of it. But whether this be the right way, whereby the universal medicine of the ancient Philosophers (by whose means all metals can be changed or transmuted into gold) is to be attained unto, I will not dispute; yet I believe that peradventure there may be another subject, endued with a far higher Tincture than gold is, which obtained no more from nature, then it does need itself for its own fixedness. However, we may safely believe, that a true Anima or Tincture of gold, if it be well separated from its impure black body, may be exalted and improved in color; so that afterwards of an imperfect body a greater quantity, than that was from which it was abstracted, may be improved and brought to the perfection of gold. But waving all this, it is true and certain, that if the gold be deprived of its Tincture, the remaining body can no more be gold; as is demonstrated more at large in my Treatise (de Auro potablili vero) of the true potable Gold: And this I mentioned here only therefore, that in case the lover of this Art, in his work should meet perchance with such a white grain, he may know, from whence it does proceed.

I could have forborn to set down the preparation of the fulminating gold, and so save paper and time, in regard that it Is described by others: but because I promised in the first part to teach how, to make the flores of gold, and that those are to be made out of fulminating (or thundering gold), I thought it not amiss to describe its preparation, that the lover of this Art need not first have his recourse to another book for to find out the preparation, but by this my book may be finished with a perfect Instruction for the making of the flores of gold, and this is the common way for to make AURUM FULMINANS, known unto most Chymists; but in regard that easily an error may be committed in it, either by pouring on too much of the liquor of Tartar (especially when it is not pure enough), so that not all the gold does precipitate, but part of it remains in the solution, whereby you would have loss; or else, the gold falling or precipitating into a heavy calx, which does not fulminate well, and is unfit for to be sublimed into flores.

Therefore I will here set down another and much better way, whereby the gold may be precipitated quite and clean out of the AQUA REGIA without the least loss, and so that it comes to be very light and yellow, and does fulminate twice as strong as the former, and there is no other difference between this and the former preparation, but only that instead of the oil of Tartar, you take the spirit of urine, or of salt armoniack for to precipitate the dissolved gold thereby; and the gold (as before said) will be precipitated much purer, than it is done by the liquor of the salt of Tartar, and being precipitated, it is to be edulcorated and dried, as above in the first preparation has been taught.

The Use of Aurum Fulminans.

There is little to write of the use of AURUM FULMINANS in physick; for because it is not unlocked, but is only a gross calx and not acceptable to the nature of man, it can do no miracle. And although it be used to be given PER SE from 6, 8, 12, grains to 1/24 ounce for to provoke sweating in the Plague, and other malignant fevers, yet it would never succeed so well as was expected. Some have mixed it with the like weight of common sulphur, and made it red hot (or calcined it) whereby they deprived it of its fulminating virtue, supposing thus to get a better medicine, but all in vain, for the gold calx would not be amended by such a gross preparation. But how to prepare a good medicine out of AURUM FULMINANS, so that it may be evidently seen, that the gold is no dead body, nor unfit for physick, but that it may be made quick and fit for to put forth or show forth those virtues which it pleased God to treasure up in it, I shall here briefly discover.

    First, get such an instrument (as above has been taught) made for you out of Copper, but not too big, nor with a lid at the top, but only with a pipe, unto which a receiver may be applied, which must not be luted to it; but it suffices, that the pipe enters far into the belly of the receiver; and at the lower part it must have a flat bottom, that it may be able to stand: over the bottom there must be a little hole with a little door, that closes very exactly: and there must be also two little plates or scales of silver or copper, as big as the nail of one's finger, whereupon the AURUM FULMINANS is to be set into the Instrument; which is to stand upon a Trivet, under which you are to lay some burning Coals for to warm or heat the bottom withal. The Instrument together with the glass Receiver being so ordered, that it stands fast and also the bottom thereof being warmed or heated, then with little pincers one of the little scales, containing 2, 3, or 4 grains of AURUM FULMINANS must be conveyed upon the Instrument set upon the warm bottom, and then shut the little door, and when the gold does feel the heat, it kindles and gives a clap, and there is caused a separation, and especial unlocking of the gold; for as soon as the clap is gone, the gold does go through the pipe like a purple colored smoke into the receiver, and sticks on everywhere like a purple colored powder. When the smoke is vanished, which is soon done, then take the empty scale out of the Instrument or Oven, and set it with the gold, which will likewise fulminate and yield its flores. Then the first being cooled in the meantime, is to be filled again and put in, instead of that which is empty, and taken out, putting in one scale after another by turns, continue it so long till you have got flores enough: After the sublimation is performed, let the Copper Vessel grow cold, and then sweep or brush the gold powder which is not sublimed with a hares foot, or goose feather out of the vessel, which powder serves for nothing, but to be melted with a little borax, and it will be good gold again, but only somewhat paler than it was before it was made into fulminating gold. But the flores in the receiver cannot be brushed out thus, especially when they are cast in with an addition of salt Nitre, as by the flores of silver hereafter shall be taught, because they are something moist, and therefore pour in as much of dephlegmed Tartarised spirit of wine unto it, as you think to be enough, for to wash off the flores with. This done, pour out the spirit of wine, together with the burnt Phoenix, into a clean glass, with a short neck, set it (being well luted first) into a gentle Balneum, or into warm ashes for some days, and the spirit of wine in the meantime will be colored with a fair red, which you must pour off and then pour on other fresh spirit and set it in a warm place for to be dissolved, this being likewise colored, put both the extracts together in a little glass body, and abstract the spirit of wine (in Balneo) from the Tincture, which will be a little in quantity, but of a high red color and pleasant in taste. The remaining flores from which the Tincture is extracted, may be with water washed out of the glass, and then dried if they are to be melted; and they will yield a little pale gold, and the most part turns into a brown glass, out of which perchance something else that is good may be made, but unknown to me as yet.

    N. B. if you mix the AURUM FULMINANS with some salt nitre, before fulmination, then the flores will be the more soluble, so that they yield their Tincture sooner and more freely, than alone of themselves; and if you please, you may add thereto thrice as much salt nitre, and so sublime them in flores, in the same manner, as shall be taught for the making of the flores of silver.

The Use of the Tincture of Gold.

    The extracted Tincture is one of the best of those medicines, which comfort & cheer up the heart of man, renew and restore to youthfulness, and cleanse the impure blood in the whole body, whereby many horrible diseases, as the leprosy, the pox, and like may be rooted out.

    But whether this Tincture by the help of fire may be further advanced into a fixed substance I do not know; for I have proceeded farther in it, than here is mentioned.

Of the Flores of Silver and of its Medicine.

    Having promised in the first part of this book (when I was describing the preparation of flores out of Metals) to teach in the second part to make the flores of gold and silver, those of gold being dispatched; there follows now in order after the gold, to speak also of silver and of its preparation, which is to be thus performed.

    Take of thin laminated or small granulated fine silver as much as you please, put it into a little separating glass body, and pour upon it twice as much in weight of rectified spirit of salt nitre, and the spirit of salt nitre will presently begin to work upon the silver and to dissolve it. But when it will not dissolve any more in the cold, then you must put the glass body into warm sand or ashes, and the water will presently begin to work again; let the glass stand in the warm ashes, until all the silver be dissolved. Then put the solution out of the little glass body, into another such as is cut off at the top, and put on a little head or Limbeck, and in sand abstract the moity[10] of the spirit of salt nitre from the dissolved silver; then let the glass body remain in the sand till it be cool; after take it out, and let it rest for a day and a night, and the silver will turn into white foliated crystals, from which you must pour off the remaining solution which is not turned; and from thence abstract again the moity of the spirit, and let it shoot or turn in a cold place; and this abstracting and crystalizing you are to reiterate, until almost all the silver is turned to Crystals; which you must take out and lay upon filtering paper to dry, and so keep it for such further use, as hereafter shall be taught. The remaining solution, which is not crystallized, you may in a copper vessel by adding of sweet water thereto, precipitate over the fire into a calx, and then edulcorate and dry it, and keep it for other use, or else melt it again into a body. Or else you may precipitate the same with salt water, and so edulcorate and dry it; and you will have a calx, which does melt by a gentle fire, and is of a special nature, and in the spirit of urine, of salt Armoniack, of Harts—horn, of Amber, of Soot, and of hair it does easily dissolve; and it may be

---

[10] "Moity" sometimes refers to a foreign particle. -PNW

prepared or turned into good medicines, as shortly in our treating of the spirit of urine shall be taught. Or else, you may choose not to precipitate the remaining solution of silver, but with the spirit of urine to extract an excellent Tincture, as hereafter shall be taught.

Of the Use of the Crystals of Silver.

These crystals may be safely be used in Physick alone by themselves 3, 6, 9, 12 grains thereof being mixed with a little sugar, or else made up into pills; they do purge very gently and without danger; but by reason of their bitterness they are somewhat untoothsome to take; also, if they are not made up into pills, they color the lips, tongue and mouth quite black (but the reason of that blackness belongs not to this place to treat of, but shall by and by follow hereafter). Also, if they touch metals, as Silver, Copper and Tin, they make them black and ugly, and therefore they are not much used. But if you put into the solution of silver (before it is reduced into Crystals) half as much quicksilver as there was of the silver, and so dissolve them together and afterwards let them shoot together, there will come forth very fair little square stones like unto Allome, which do not melt in the air, as the former foliated ones use to do; neither are so bitter, and they purge also quicker and better, than those that are made only of silver.

How to Sublime the Crystals of Silver into Flores, and then to make a good Medicine of the Flores.

Take of the aforesaid Crystals of Silver as many as you please, and upon a grinding stone made warm first, grind as much purified and well dried salt nitre amongst it, then put into your Iron distilling vessel (to the pipe whereof there is to be applied and luted a great receiver) coals made into powder two inches high, and make a fire under it, that the vessel everywhere together with the coals that are in it, become red hot. Then take off the lid, and with a ladle throw in at once of your Crystals of silver 1/8 ounce more or less, according as you think that your receiver in regard of its bigness is able to bear. This done, presently put on the lid, and the Salt nitre together with the crystals of silver will be kindled by the coals that lie on the bottom of the vessel, and there will come forth a white silver fume through the pipe into the receiver, and after a while when the cloud is vanished in the receiver, cast in more, and continue this so long, and until all your prepared silver is cast in; then let it cool, and take off the receiver, and pour into it good Alcolized spirit of wine, and wash the flores with it out of the receiver, and proceed further with them, as above you have been taught to proceed with the gold, and you will get a greenish liquor, which is very good for the brain.

    Take the coals out of the distilling vessel, and make them into fine powder, and wash them out with water, to the end that the light coal dust may be got from it, and you will find much silver dust (or a great many little silver grains) which the salt nitre could not force over, which you may reduce, for it will be good silver.

    There may also be made a very good medicine out of the crystals of silver, which will be little inferior to the former, whereby the diseases and infirmities of the brain may be very well remedied, which is done thus.

How to Make a Green Oil out of Silver.

    Pour upon Crystals of silver twice or thrice as much (in weight) of the strongest spirit of Salt Armoniack, put it in a glass with a long neck well closed, into a very gentle warmth for the space of 8 or 14 days in digestion, and the spirit of salt Armoniack will be tinged with a very fair blue color from the silver, then pour it off, and Filter it through brown paper, and then put it in a little glass retort or glass body, and abstract in Balneo by a gentle fire, almost all the spirit of salt Armoniack (which is still good for use) and there will remain in the bottom a grass-green Liquor, which is to be kept for a Medicine.

    But in case that you should miss, and abstract too much of the spirit from the Tincture of silver, so that the Tincture be quite dry and turned to a green salt, then you must pour upon it again as much of the Spirit of Salt Armoniack, as will dissolve the green salt again to a green Liquor, but if you desire to have the Tincture purer yet, then abstract all moistness from it, to a stony dryness: upon which you must pour good spirit of Wine, which will quickly dissolve the stone, and then Filter it, and there will remain feces, and the Tincture will be fairer: from which you must abstract most of the spirit of wine, and the Tincture will be so much the higher in virtue. But if you please, you may distil that green salt or stone (before it be extracted again with the spirit of wine) in a little glass retort, and you will get a subtle spirit and a sharp oil, and in the bottom of the retort there remains a very fusile silver which could not come over.

It is to be admired, that when you pour spirit of salt Armoniack, or spirit of wine upon that stone, for to dissolve it, that the glass comes to be so cold by it, that you hardly are able to endure it in your hand, which coldness in my opinion comes from the silver (being so well unlock) which naturally is cold.

The Use of the Green Liquor in Alchemy, and for Mechanical Operations.

This green Liquor serves not only for a medicine, but also for other Chymical operations (for both Copper and glass may be easily and very fairly silvered over therewith) very useful for those that are curious and love to make a show, with fair household-stuff; for if you get dishes, treacher-plates, salters, cups and other vessels made of glass, after the same fashion as those of silver use to be made, you may easily and without any considerable charge silver them over therewith within and without, so that by the eye they cannot be discerned from true silver plate.

Besides the above-related good Medicines, there may be made another and especial good one out of the crystals of silver, viz. dissolving and digesting them (for a space of time) with the universal water, which has been distilled by nature itself; and is known to everybody; and after its digesting for a short time, and changing into several colors, there will be found a pleasant essence, which is not so bitter as the above-described green liquor, which is not brought yet by heat to ripeness and maturation.

N. B. In this sweet universal Menstruum, may also all other metals by a small heat and the digestion of a long time be ripened and fitted for Medicines (having first been reduced into their vitriols and salts) and then they are no more dead bodies, but by this preparation have recovered a new Life, and are no more the metals of the covetous, but may be called the metals of the Philosophers, and of the Physicians.

Besides Physick or Physical Use.

Lastly, there may be many pretty things more effected (besides the medicinal use) by means of the Crystals of silver, viz. when you dissolve them in ordinary sweet rain water, you may dye beards, hair, skin, and nails of men or beasts into carnation or pink red, brown and black, according as you have put more or less thereof in the water; or else, according as the hair was more or less times wetted therewith, whereby the aspect of Man and Beast (which sometimes in several occasions may not be contemned) is changed, so that they cannot be known.

This coloring or dye may be also performed with Lead or Mercury no less than with silver, but otherwise prepared, whereof in the fourth part.

Now I have taught how to make flores and tinctures of gold and silver by help of the acid spirit of Nitre. There may be many other medicines taught to be made from them, but in regard that they belong not to this place, they shall be reserved for other places of the second and also for the other following parts.

As by the help of the spirit of Nitre, good Medicines may be made from gold and silver, so the like may be done out of other inferior metals. But in regard that their description is fitter for other places of this Book, I omit them here. Yet nevertheless, I thought good to describe one preparation of every metal; after silver therefore follows now Copper.

A Medicine out of Copper Externally to be Used.

Dissolve burnt plates of Copper in spirit of salt, and abstract the spirit again from thence to a dryness, but not too hard, and there will a green mass remain behind, which you may cast in by little and little, and so distil it, as of silver has been taught. It does yield a strong and powerful spirit, and flores also for outward use in putrid wounds, to lay a good ground thereby for the healing.

A Medicine out of Iron and Steel.

In the same manner, you may proceed with iron and steel, and there will remain behind a good crocus of a great stipticity or astringency, especially out of iron or steel, and may with good success be mixed with ointments and plaisters.

Of Tin and Lead.

If Tin or Lead be dissolved therein, after the abstracting of part of the spirit, they will shoot into clear and sweet crystals. But Tin is not so easily dissolved as Lead; both may safely be used for medicines. Also, there may be spirits and flores got out of them by distilling. The rehearsing of the Preparation is needless, for what for the preparing of silver has been taught, is to be understood also of other metals.

The Use of the Crystals of Lead and Tin.

The Crystals of Lead are admirably good to be used in the plague for to provoke sweating and expel the venome out of the body; they may also with credit be used in the bloody flux. Externally dissolved in water, and clothes dipped therein and applied, they excellently cool and quench all inflammations, in what part of the body so ever they do befall. Likewise, the spirit thereof used per se (and the flores mixed among ointments) do their part sufficiently.

But the crystals of Tin do not prove altogether so quick in operation, though they do act their part also, and they are more pleasant than those that are made of Lead; for in Tin there is found a pure sulphur of gold; but in Lead a white sulphur of silver, as is proved in my TREATISE OF THE GENERATION AND NATURE OF METALS.

Of Mercury.

When you dissolve common Mercury in rectified spirit of Nitre, and abstract the spirit from it again, then there will remain behind a fair red glistering precipitate; but when the spirit is not rectified, it will not be so fair, because that the impurity of the spirit remains with the Mercury and pollutes it. This calcined Mercury is called by some MERCURIUS PRAECIPITATUS, and by others TURBITH MINERALE, wherewith the Surgeons, and sometimes other unskillful Physicians do cure the Pox; they give at once 6, 8, 10 grains, (more or less) according to its preparation and force in operation to the patient; for if the spirit be not too much abstracted from it, it works much stronger, than by when a strong fire it is quite separated from it; for the spirits that remain with the Mercury make it quick and active, which else without the spirits would not be such.

    The other metals also, if they be not first made soluble by salts or spirits, can perform either none of but very small operation, unless it be Zink or Iron, which being easily soluble, can work without any foregoing dissolution, as has been shown above, when we treated of the oil of vitriol. But that the sharp spirits are the cause of that operation, may hence be perceived, and made manifest that although you take 1/4 ounce of quick-silver, and pour it down into the stomach, yet it would run out again beneath, as above it was poured in. But if it be prepared with spirits of salt, then but few grains of it will work strongly, and the more it is made soluble, the stronger it works; as you may see when it is sublimed from salt and vitriol, that it grows so strong thereby, that one grain does work more than eight or ten grains of Turbith Mineral, and three or four grains thereof would kill a man, because of its mighty strength. Also, it works extremely, and much more than the sublimate, when it is dissolved in spirit of Nitre and crystalized, so that you cannot well take it upon your tongue without danger: Which some perceiving, evaporate the AQUA FORTIS by a gentle heat from it, so that the MERCURIUS remained yellow, which in a smaller dose wrought more than the red, from which the spirits were quite evaporated. And they used it only externally, strewing it into impure sores, for to corrode or fret away the proud flesh, not without great pain to the patient: but also without distinction of young or old, gave it inwardly for to purge; which is one of the most hurtful Purges that can be used. For this evil guest, however he be prepared, cannot leave his tricks, unless he be reduced into such a substance, as that it never can be brought back to a running Mercury, for then much good can be done in physick without any hurt or prejudice to the health of man, whereof perchance something more shall be said in another place.

    I cannot omit for the benefit of young innocent Children, to discover a great abuse. For it is grown very common almost among all that deal in physick, that as soon as a little child is not well before they know whether it will be troubled with worms, or with anything else, they presently fall upon Mercury, supposing that in regard it has no taste, it is so much the better for to get the Children to take it for to kill the Worms.

    But those men do not know the hurtful nature of it, which it does show against the sinews and Nerves. For some are of opinion, that if they know to prepare Mercury so, that it can be given in a greater dose (as is to be seen in sublimed MERCURIUS DULCIS) that then it is excellently prepared: but they are in a great error, and it were much better, it was not so well prepared, that the less hurt might be done to Man, in regard that then they durst not give it in so great a dose. For if that which is prepared with AQUA FORTIS or spirit of salt nitre be used in the pox to men that are advanced in years, it cannot do so much hurt, because it is given in a small dose, and does work with them, whereby nature gets help for to overcome and expel that hurtful venome, and its malignity is abated by the strong salivation, which provident nature has planted in it, so that not so much mischief can come by it, as by MERCURIUS DULCIS, whereof is given to little weak Children from ten to thirty grains at once, which commonly (unless they be of a strong nature, and do grow out of it) does cause a weakness and lameness in their limbs, so that (if they do not come to be quite lame at last) they have a long time to struggle withal, till they overcome it.

In like manner those also do err, which do shake Mercury in water or beer so long, until the water come to be gray-colored, and so give that water or beer to little children to drink for the Worms, pretending that they do not give the substance or body of Mercury; but only its virtue; but this gross Preparation is no better than if they had ministered the running Mercury itself. Neither have I ever seen that the use of MERCURIUS DULCIS, or of the grey colored water was seconded with good success in killing of the Worms. But it is credible, that it may be done by yellow or red precipitate, in regard of its strong operation. But who would be such an Enemy to his Child, as to plague and torture it with such a hurtful and murthering medicine, especially there being other medicines to be had, which do no harm to the children, as is to be found in iron or steel, and the sweet oil of vitriol.

And so much of the abuse of Mercury: I hope it will be good warning unto many, so that they will not so easily billet such a tyrannical guest in any one's house, thereby the ruin thereof of necessity must follow. And that cure deserves no praise at all, thereby one member is cured with the hurt of two or three other members. As we see by the Pox, when one infected member is cured by Mercury, and that but half, and not firm at all, that all the rest of the body is endangered thereby for the future. And therefore, it would be much better that such crude horse-physick might be severed from good medicaments, and such used instead of them, as may firmly, safely, and without prejudice to other parts perform the cure, of which kind several are taught in this book. But in case that you have Patients, which have been spoiled by such an ill-prepared Mercury, then there is no better remedy to restore them, than by medicines made of metals, therewith Mercury has great affinity, as of gold and silver: for when they are often used, they attract the Mercury out of the members, and carry it along with them out of the body, and so do rid the body thereof. But externally the precipitated Mercury may more safely be used, than internally, in case there be nothing else to be had, viz. to corrode or eat away the proud flesh out of a wound. But if instead of it there should be used the corrosive oil of Antimony, Vitriol, Allome or common salt it would be better, and the cure much the speedier; and it would be better yet, that in the beginning good medicaments were used to fresh wounds, and not by carelessness to reduce them to that ill condition, that afterwards by painful corrosives they must be taken away. But such a Mercury would serve best of all for soulders, beggars, and children that go to school; for if it be strewed upon the head of children, or into their clothes, no louse will abide there any longer. In which case Mercury must by his preparation not be made red, but only yellow, and it must be used warily, and not be strewed on too thick, lest the flesh be corroded, which would be the occasion of great mischief.

Of Aqua Fortis.

Out of Salt nitre and vitriol, taking of each a like quantity (for if the water is to be not altogether so strong) two parts of vitriol to one part of salt nitre, a water distilled is good to dissolve metals therewith, and to separate them from one another; as gold from silver, and silver from gold, which is in the fourth part punctually shall be taught.

The AQUA FORTIS serves also for many other Chymical operations to dissolve and fit metals thereby, that they may be reduced the easier into medicaments: but because the spirit of salt nitre and AQUA FORTIS are almost all one, and have like operations: for if the AQUA FORTIS be dephlegmed and rectified, you may perform the same operations with it, which possibly may be performed with the spirit of salt nitre; and on the other side the spirit of salt nitre will do all that can be done with the AQUA FORTIS, whereof in the fourth part shall be spoken more at large.

Now I know well that ignorant laborers (which do all their work according to custom), without diving any further into the Nature of things, will count me a Heretic (because I teach, that the AQUA FORTIS made of vitriol and salt nitre is of the same nature and condition with the spirit of salt nitre, which is made without vitriol) saying that the AQUA FORTIS does partake likewise of the spirit of vitriol, because vitriol also is used in the preparation of it. To which I answer, that although vitriol be used in the preparation of it, yet for all that in the distilling, nothing or but very little of its spirit comes over with the spirit of salt nitre, and that by so small a heat it cannot rise so high, as the spirit of salt nitre does: and the vitriol is added only therefore unto the salt nitre that it may hinder its melting together, and so the more facilitate its going into a spirit. And the more to be convinced of this truth, the unbelieving may add to such spirit of salt nitre, as is made by itself, a little of oil of vitriol likewise made by itself, and try to dissolve silver guilded with it, and he will find that his spirit of salt nitre by the spirit of vitriol is made unfit to make a separation; for it preys notably upon the gold, which is not done by AQUA FORTIS.

Of the Sulphurised Spirit of Salt Nitre.

There may also be made a spirit of salt nitre with sulphur, which is still in use with many, viz. they take a strong earthen retort, which has a pipe at the top, and fasten it into a furnace, and having put salt nitre into it, they let it melt, and then through the pipe they throw pieces of sulphur of the bigness of a pea, one after another, which being kindled, together with the nitre does yield a spirit called by some spirit of salt nitre, and by others oil of sulphur, but falsely; for it is neither of both, in regard that metals cannot be dissolved therewith as they are done with other spirit of salt nitre or sulphur; neither is there any great use for it in physick, and if it were good for any Chymical operations, by the help of my distilling instrument might easily be made and in great quantity.

N. B. But if salt nitre be mixed with sulphur in due proportion, and in the first furnace be cast upon quick coals, then all will be burnt, and a strong spirit comes over, whose virtue is needless here to describe; but more shall be mentioned of it in another place.

Of the Clissus.

Among the Physicians of this latter age, there is mention made of another spirit, which they make of Antimony, Sulphur, and salt nitre, a like quantity taken of each, which they call CLISSUS, and which they have in high esteem, and not without cause, because it can do much good if it be well prepared.

The inventor, for the making thereof used a retort with a pipe, as was mentioned by the sulphurized spirit of salt nitre, through which pipe he threw in his mixture. And it is a good way if no better be known: but if the Author had known my invention and way of distilling, I doubt not but he would have set aside his, that has a nose or pipe retort, and made use of mine.

The materials indeed are good, but not the weight or proportion; for to what purpose so great a quantity of sulphur, it being not able to burn away all with so small a quantity of salt nitre. And if it does not burn away, but only sublime & stop the neck of the retort, whereby the distillation is hindered, how can it then yield any virtue? Therefore, you ought to take not so much sulphur, but only such a quantity as will serve to kindle the salt nitre, viz. to 1 lb. of salt nitre four drams of sulphur: but because Antimony also is one of the ingredients, which has likewise much sulphur (for there is no Antimony so pure, but it contain much combustible sulphur, as in the fourth part of this book shall be proved), therefore it is needless to add so much sulphur unto Antimony, to make it burn, because it has enough of itself. And therefore, I will set down my composition, which I found to be better than the first.

Take Antimony 1 lb., salt nitre 1/2 lb., sulphur 3 ounces; the materials must be made into small powder and well mixed, and at once cast in 3 ounces thereof, and there will come over a sulphureous acid spirit of Antimony, which will mix itself with the water, which has been put before in the receiver; which after the distillation is finished must be taken out and kept close for its use. It is a very good diaphoretick (for sweat provoking) medicine especially in fevers, the plague, epilepsie, and all other diseases, whose cure must be performed by sweating. The CAPUT MORTUUM may be sublimed into flores in that furnace, which is described in the first part.

Of the Tartarised Spirit of Nitre.

In the very same manner there may also be distilled a good sweat-provoking spirit out of salt nitre and Tartar, a like quantity taken of each, which is very good to be used in the plague and malignant fevers.

The CAPUT MORTUUM is a good melting powder for to reduce the calxes of metals therewith; or else you may let it dissolve in a moist place to oil of Tartar.

Of the Tartarised Spirit of Antimony.

A much better spirit yet may be made of Tartar, salt nitre, and Antimony, a like quantity being taken of each, and made into fine powder, and mixed well together, which though it be not so pleasant to take, is therefore not to be despised. For not only in the plague and fevers, but also in all obstructions and corruptions of blood it may be used with admiration of its speedy help.

The CAPUT MORTUUM may be taken out, and melted in a crucible, and it will yield a REGULUS, the use whereof is described in the fourth part.

Out of the scoria or dross a red Tincture may be extracted with spirit of wine, which is very useful in many diseases. But before you extract with spirit of wine, you may get a red lixivium out of it with sweet water, which lixivium may be used externally for to mend the faults of the skin and to free it from scabbiness.

Upon this lixivium if you pour Vinegar or any other acid spirit, there will precipitate a red powder, which if it be edulcorated and dried may be used in physick. It is called by some sulphur AURATUM DIAPHORETICUM: but it is no Diaphoretick, but makes strong vomits, and so in case of necessity, when you have no better medicine at hand, it may be used for a vomitory from 6, 7, 9, 10, 15.

Also, out of the scoria there may be extracted a fair Sulphur with the spirit of urine and distilled over the Limbeck, which is very good for all diseases of the lungs.

Of Stone—coles.

If you mix stone—coals with a like quantity of salt nitre, and distill them, you will get an admirable spirit and good to be used outwardly; for it cleanses and consolidates wounds exceedingly, and there will also come over a metallical virtue in the form of a red powder, which must be separated from the spirit, and kept for its use. But if you cast in stone—coals alone by themselves, and distill them, there will come over not only a sharp spirit, but also a hot and blood red oil, which does powerfully dry and heal all running ulcers; especially it will heal a scald head better than any other medicine, and it does consume also all moist and spongious excrescencies in the skin, where ever they be: but if you sublime stone—coles in the furnace described in the first part, there comes over an acid metallical spirit, and a great deal of black light flores, which suddenly stanch bleeding, and used in plaisters, are as good as other metallical flores.

Of the Sulphureous Spirit of Salt Nitre or Aqua Fortis.

If you take one part of sulphur, two parts of nitre, and three parts of vitriol, and distill them, you will get a graduating AQUA FORTIS, which smells strongly of sulphur; for the sulphur is made volatile by the salt nitre and vitriol. It is better for separating of metals, than the common AQUA FORTIS.

If silver be put in, it grows black, but not fixed; some of it poured into a solution of silver a great deal of black calx will precipitate, but does not abide the trial. You may also abstract a strong SULPHUREOUS volatile spirit from it, which has like virtue as well internally as externally for bathes, and may be used like unto a volatile spirit of Vitriol or Allome.

Of the Nitrous Spirit of Arsenick.

If you take white Arsenick and pure salt nitre of each a like quantity ground into fine powder, and distill them, you will get a blue spirit, which is very strong, but no water must be put into the receiver, else it would turn white, for Arsenick, from which the blue comes, is precipitated by the water. This spirit dissolves and graduates copper as white as silver, and makes it malleable but not fixed. The remaining CAPUT MORTUUM makes the copper white, if it be cemented therewith, but very brittle and unmalleable, but how to get good silver out of Arsenick and with profit, you shall find in the fourth part. In physick the blue spirit serves for all corroding cancerous sores, which if they be anointed therewith, will be killed thereby, and made fit for healing.

To Make a Spirit of Sulphur, Crude Tartar and Salt nitre.

If you grind together one part of Sulphur, two parts of Crude Tartar, and four parts of salt nitre, and distill it Philosopher—like, you will get a most admirable spirit, which can play his part both in Physick and Alchymy, I will not advise anybody to distill it in a retort; for this mixture, if it grows warm from beneath, it fulminates like Gunpowder; but if it be kindled from above, it does not fulminate, but only burns away like a quick fire: metals may be melted and reduced thereby.

To Make a Spirit out of Salt of Tartar, Sulphur, and Salt—nitre.

If you take one part of Salt of Tartar, and one part and a half of Sulphur, with three parts of salt nitre, and grind them together, you will have a composition, which fulminates like AURUM FULMINANS, and the same also (after the same manner as above has been taught with gold) may be distilled into flores and spirits, which are not without special Virtue and Operation. For the corruption of one thing is the generation of another.

How to Make a Spirit of Saw-dust, Sulphur and Salt Nitre.

If you make a mixture of one part of Saw-dust made of Tilia or Linden-wood, and two parts of good sulphur, and nine parts of purified and well dried salt nitre, and cast it in by little and little, there will come over an acid spirit, which may be used outwardly, for to cleanse wounds that are unclean. But if you mix with this composition minerals or metals made into fine powder, and then cast it in and distill it, there will come not only a powerful metallical spirit, but also a good quantity of flores, according to the nature of the mineral, which are of no small virtue: for the minerals and metals are by this quick fire destroyed and reduced to a better condition, whereof many things might be written: but it is not good to reveal all things. Consider this sentence of the Philosophers. IT IS IMPOSSIBLE TO DESTROY WITHOUT A FLAME, THE COMBUSTIBLE SULPHUR OF THE CALX, WHICH THE DIGGED MINE DOES DO.

Also, fusible minerals and metals may not only be melted, therewith, but also cupellated in a moment upon a Table, in the hand or in a nutshell; whereby singular proofs of ores and metals may be made, and much better, than upon a Cupel, whereof further in the fourth part of this book. Here is opened unto us a gate to high things; If entrance be granted unto us, we shall need no more books to look for the Art in them.

To Make Metallical Spirits and Flores by the Help of Salt-nitre and Linen Cloth.

If metals be dissolved in their appropriated Menstruums, and in the solution (wherein a due proportion of salt nitre must be dissolved) fine linen rags be dipped and dried, you have a prepared metal, which may be kindled, and (as it was mentioned above concerning the saw-dust) through the burning away and consuming of their superfluous sulphur, the mercurial substance of the metal is manifested. And after the distillation is ended, you will find a singular purified calx, which by rubbing colors other metals, as that of gold guilds silver, that of silver silvereth over copper, and copper calx makes iron look like copper, & etc. which coloring though it cannot bring any great profit, yet at least for to show, the possibility, I thought it not amiss to describe it; and perchance something more may be hid in it, which is not given to everyone to know.

Of Gunpowder.

Of this mischievous composition and diabolical abuse of Gunpowder much might be written: but because this present world takes only delight in shedding innocent blood, and cannot endure that unrighteous things should be reproved, & good things praised, therefore it is best to be silent, and to let everyone answer for himself, when the time comes that we shall give an account of our steward-ship, which perhaps is not far off; and then there will be made a separation of good and bad, by him that tries the heart, even as gold is refined in the fire from its dross. And then it will be seen what Christians we have been. We do all bear the name, but do not prove ourselves to be such by our works; everyone thinks himself better than others, and for a words sake which one understands otherwise, or takes in another sense than the other (and though it be no point, whereon salvation does depend) one curses and condemns another and persecutes one another unto death, which Christ never taught us to do, but rather did earnestly command us that we should love one another, reward evil with good, and not good with evil, as nowadays everywhere they use to do; every one stands upon his reputation, but the honor of God and his command are in no repute, but are trampled underfoot, and Lucifer's pride, vain ambition, and Pharisaical hypocrisie or show of holiness; has so far got the upper hand with the learned, that none will leave his contumacy or stubbornness, or recede a little from his opinion, although the whole world should be turned upside down thereby. Are not these fine Christians? By their fruit, you shall know them, and not by their words. Wolves are now clothed with sheep skins, so that none of them almost are to be found, and yet the deeds and works of wolves are everywhere extant.

    All good manners are turned into bad, women turn men, and men women in their fashion and behavior, contrary to the institution and ordinance of God and Nature. In brief, the world goes on crutches. If HERACLITUS and DEMOCRITUS should now behold this present world, they would find exceeding great cause for their lamenting and laughing at it. And therefore, it is no marvel, that God sent such a terrible scourge as gun-powder upon us; and it is credible, that if this does not cause our amendment, that a worse thing will follow, viz. thunder and lightning falling from Heaven, whereby the world shall be turned upside down, to make an end of all pride, self-love, ambition, deceit and vanity. For which the whole Creation does wait, fervently desiring to be delivered from the bondage thereof.

Now this preparation, which is the most hurtful poison, a terror unto all the living, is nothing else but a FULMEN TERRESTRE denouncing unto us the wrath and coming of the Lord. For Christ to judge the world is to come with thundering and lightning: and this earthly thunder perchance is given us to put us in mind and fear of that which is to come, but this is not so much as thought on by men, who prepare it only for to plague and destroy mankind therewith in a most cruel and abominable manner, as everyone knows.

For none can deny but that there is no nimbler poison, than this gun-powder. It is written of the Basiliske, that he kills man only by his look, which a man may avoid, and there are but few (if any at all) of them found: but this poison is now prepared and found everywhere.

How often does it fall out, that a place wherein this powder is kept is stricken with thunder as with its like, in so much that all things above it are in a moment destroyed, and carried up into the air? Also in sieges, when an Ordnance is discharged, or Mines blown up, all whom it lays hold on, are suddenly killed, and most miserably destroyed. What nimbler poison then could there be invented? I believe there is none, who will not acknowledge it to be such.

And seeing that the ancient Philosophers and Chymists were always of opinion, that the greater the poison is, the better medicine may be made of it, after it is freed from the poison, which with us their posterity is proved true by many experiences;

as we see by Antimony, Arsenic, Mercury, and the like minerals, which without preparations are mere poison, but by due preparation may be turned into the best and most effectual medicaments, which though not everyone can comprehend or believe, yet Chymists know it to be true, and the doing of it is no new thing to them. And because I treat in this second part of medicinal spirits, and other good medicaments, and finding that this which can be made from gunpowder, is none of the least, I would not omit in some measure, and as far as lawfully may be done, to set down its preparations: which is thus performed.

How to Make a Spirit of Gunpowder.

Your distilling vessel being made warm, and a great receiver with sweet water in it being applied to it without luting, put a dish with gunpowder, containing about 12 or 15 grains a piece, one after another into it; in the same manner as above was taught to do with gold. For if you should put in too much of it at once, it would cause too much wind and break the receiver.

As soon as you have conveyed it into the vessel, shut the door, and the gunpowder will kindle, and give a blast that it makes the receiver stir, and a white mist or steam will come over into the receiver. As soon as the powder is burnt, you may cast in more before the mist is settled, because else the distilling of it would cost too much time, and so you may continue to do until you have spirit enough. Then let the fire go out, and the furnace grow cool, and then take off the receiver, pour the spirit with the water that was poured in before (the flores being first everywhere washed off with it) out of the receiver into a glass body, and rectify it in a B.[11] through a limbick, and there will come over a muddy water, tasting and smelling of sulphur: which you must keep. In the glass body, you will find a white salt, which you are to keep likewise. Take out the CAPUT MORTUUM, which remained in the distilling vessel, and looks like gray-salt, calcine it in a covered crucible, that it turns white, but not that it melts; and upon this burnt or calcined salt, pour your stinking water, which came over through the Limbeck, and dissolve the calcined white salt with it, and the feces which will not dissolve cast away. Filter the solution, and pour it upon the white salt, which remained in the glass body, from which the sulphureous spirit was abstracted before, and put the glass body (with a limbeck luted upon it) into sand, and abstract the sulphureous water from it, which will be yellowish, and smell more of sulphur than it did before. This water if it be abstracted from the salt several times, will turn white, almost like unto milk, and taste no more of sulphur, but be pleasant and sweet. It is very good for the diseases of the lungs. Also, it does guild silver, being anointed therewith, although not firmly, and by digestion it may be ripened and reduced into a better medicine.

The salt which remained in the glass body, urge with a strong fire, such as will make the sand, wherein the glass stands red hot, and there will sublime a white salt into the limbeck, in taste almost like unto salt Armoniack, but in the midst of the glass body, you will find another, which is yellowish, of a mineral taste and very hot upon the tongue.

---

[11] Balneum Maria…Mary's Bath…Water bath…a double boiler. -PNW

The sublimed salts, as well the white which did ascend into the limbeck, as the yellow, which remained in the glass body are good to be used in the plague, malignant fevers and other diseases, where sweating is required; for they do mightily provoke sweating, they comfort and do cleanse the stomach, and cause sometimes gentle stools.

But what further may be done in Physick with it, I do not know yet.

In Alchymy it is also of use, which does not belong to this place. Upon the remaining salt, which did not sublime you may pour rain water, and dissolve it there in the glass body, (if it be whole still) else if it be broken, you may take out the salt dry, and dissolve and Filter and coagulate it again, and there will be separated a great deal of feces. This purified salt, which will look yellowish, melt in a covered crucible, and it will turn quite blood red, and as hot as fire upon the tongue, which with fresh water you must dissolve again, and then Filter and coagulate; by which operation it will be made pure and clear, and the solution is quite green before it be coagulated, and as fiery as the red salt was before its dissolution.

This grass green solution being coagulated again into a red fiery salt, may be melted again in a clean and strong crucible, and it will be much more red and fiery.

N. B. And it is to be admired that in the melting of it many fire sparks do fly from it, which do not kindle or take fire, as other sparks of coals or wood use to do. This well purified red salt being laid in a cold and moist place, will dissolve into a blood red oil, which in digestion dissolves gold and leaves the silver: this solution may be coagulated, and kept for use in Alchemy.

There may be also a precious Tincture be extracted out of it with alcolized spirit of wine, which Tincture guilds silver, but not firmly.

And as for use in Physick, it ought to be kept as a great Treasure. But if the red fiery salt be extracted with spirit of wine before gold be dissolved therewith, it will yield likewise a fair red Tincture, but not so effectual in Physick as that unto which gold is joined. And this Tincture can also further be used in Alchemy, which belongs not hither, because we only speak of medicaments.

Of the Use of the Medicine or Tincture Made of Gunpowder.

This Tincture whether with or without gold, made out of the red salt, is one of the chiefest that I know to make, if you go but rightly to work, and prepare it well; for it purifies and cleanses the blood mightily, and provokes also powerfully sweat and urine; so that it may safely and with great benefit be used in the Plague, Fevers, Epilepsy, Scurvy, in MELANCHOLIA HYPOCHONDRIACA, in the Gout, Stone, and the several kinds of them; as also in all obstructions of the Spleen and Liver, and all diseases of the Lungs, and it is to be admired that of such a hurtful thing such a good medicine can be prepared. Therefore, it would be much better to prepare good medicaments of it, to restore the poor diseased to health therewith, than to destroy with it those that are whole and sound.

I know a Chymist, that spent much time and cost to search this poisonous dragon, thinking to make the universal medicine or stone of the ancient Philosophers out of it. Especially because he saw, that so many strange changes of colors appeared, whereof mention is made by the Philosophers when they describe their medicine and the preparation thereof.

The Dragons blood, Virgins milk, Green and Red Lion, Black blacker than Black, White whiter than White and the like, more needless here to relate, which easily may persuade a credulous man as it happened also unto him. But afterward he found, that this subject in which he put so much confidence, was leprous and not pure enough, and that it is impossible to make that tingent stone of it, for to exalt men and metals, and so was glad to be contented with a good particular medicine and to commit the rest unto God.

And so much of that poisonous dragon, gunpowder: but that there is another and purer dragon, whereof the Philosophers so often made mention, I do not deny; for nature is mighty rich, and could reveal to us many arcana by Gods permission: But because we look only for great honor and riches, and neglect the poor, there is good reason why such things remain hidden from wicked and ungodly men.

To make Spirits and Flores of Nitre and Coals.

If you distill Nitre (well purified from its superfluous salt) mixed with good coals, the Egyptian Sun bird does burn away, and out of it does sweat a singular water, useful for men and metals. Its burnt ashes are like unto calcined Tartar, and for the purging of metals not to be despised.

To Make Flores and Spirits of Flints, Crystals or Sand, by Adding of Coals and Salt Nitre.

Take one part of flints or sand, and three parts of Linden coals, with six parts of good salt nitre mixed well together, and cast of it in, and the combustible sulphur of the flints will be kindled by the piercing and vehement fire of the salt nitre, and makes a separation, carrying over with it part thereof, which it turns into spirits and flores, which must be separated by filtering. The spirit tastes as if it had been made of the salt of Tartar and flints, and is of the same nature and condition; and the remaining CAPUT MORTUUM also yields such an oil or liquor in all like unto that, and therefore its condition is not described here, but you may find it where I shall treat of the spirit made of salt of tartar by adding of flints.

To Make a Spirit and Oil out of Talck with Salt Nitre.

Take one part of Talck made into fine powder, and three parts of Linden-coals, mix them with five, or six parts of good salt nitre, cast in of that mixture one spoonful after another, and there will come over a spirit and a few flores, which must be separated as has been taught above concerning flints.

The spirit is not unlike the spirit of sand: the CAPUT MORTUUM, which looks greyish, must be well calcined in a crucible, so that it melts, and then pour it out, and it will yield a white transparent mass, like flints and crystals do, which in a cold moist cellar will turn to a thick liquor, fatter in the handling than the oil of sand. It is something sharp like unto oil of Tartar; it cleanses the Skin, Hair and Nails, and makes them white; the spirit may be used inwardly to provoke sweat and urine: externally used, it cleanses wounds, and heals all manner of scabs in the body out of hand. What further may be done with it, I do not know yet: But how to bring Talck, pebbles, and the like stony things to that pass, that they may be dissolved with spirit of wine and reduced into good medicaments shall be taught in the fourth part.

To Make a Spirit, Flores, and Oil out of Tin.

If you mix two parts of the filings of Tin, with one part of good salt nitre, and cast it in, as you were taught to do with other things, then the sulphur of Tin will kindle the salt nitre, and make a flame, as if it were done with common sulphur; whereby a separation is made, so that one part of the Tin comes over in flores and spirit, and the rest stays behind, which if it be taken out, some of it in a moist place will turn into a liquor or oil, which externally may be used with good success in all ulcers for to cleanse them. It has also the virtue, if it be pertinently applied to graduate and exalt wonderfully all the colors of vegetables and animals, which would be useful for dyers. The spirit of it nightly provokes sweating: the flores being edulcorated and used in plasters, do dry and heal very speedily.

To Make a Spirit, Flores and a Liquor out of Zinck.

In like manner, as has been taught with Tin, you may also proceed with Zinck, and it will yield a good quantity of flores, and also a spirit and oil, almost of the same virtues with those made of Tin: and these flores corrected with salt nitre, are better than those which are taught to be made by themselves in the first part of the book.

To Make a Spirit, Flores and Oil of Lapis Calaminaris.

Mix two parts of salt nitre with one part of LAPIS CALAMINARIS and cast it in, and it will yield a sharp spirit very useful for separating of metals, and there will come over also a few yellow flores. The rest remaining behind is a dark green Mass very fiery upon the tongue, like salt of Tartar, and if it be dissolved with rain water, yields a grass green solution, which being not presently coagulated into salt, the green separates itself from the fixed salt nitre, and there falls to the bottom a fine red powder, and if it be edulcorated and dried, and given from one grain to ten or twelve it causes gentle stools and easy vomits, better than prepared Antimony; for LAPIS CALAMINARIS and Zinck are of the nature of Gold, as in the fourth part shall be proved: the white LIXIVIUM or lye, from which the green is precipitated, may be coagulated into white salt, like unto salt of Tartar; but if you coagulate the green solution, before the green be separated from the salt nitre, then you will get a very fair green salt, high in color and much more fiery than salt of Tartar, whereby special things may be done in Alchemy, which does not belong hither. And if you desire to make such a green salt for to use in Alchemy, you need not take so much pains, as first to distil a spirit out of the mixture, but take three or four parts of good salt nitre, and mix it with one part of LAPIS CALAMINARIS, and let this mixture boil together in a wind furnace, till the salt nitre be colored green by the LAPIS CALAMINARIS, then pour it out and separate the green goldish salt from it, and make such good use of it as you think fit.

But if you will extract a good Tincture and medicine, make it into powder, and extract it with spirit of wine, and it will yield a blood red Tincture, both in Physick and Alchemy of good use.

Further you are to take notice, that among all metals and minerals, which I know (except gold and silver) there is none found, out of which can be extracted a greenness which is of fire—proof, but only out of LAPIS CALAMINARIS, which deserves to be well considered and further thought upon.

To Make a Spirit of Salt Nitre, Sulphur and Common Salt.

Take one part of salt, two parts of sulphur, and four parts of salt nitre, grind all together, and cast in one spoonful after another to distil, and it will yield a sharp yellow spirit, which if it be put among common water, so that the water be not made too sharp of it, it is a good bath, good for many diseases; especially it heals all scabs very suddenly. The CAPUT MORTUUM may also be dissolved in water and used among bathes, and it is good likewise, but the spirit is penetrating, and does operate suddenly in shrinkings and other defects of the nerves; of such kind of bathes there shall be spoken more in the third part. Also, the remaining fixed yellow salt is good to be used in Alchemy; for it graduates silver by cementing.

To Make a Spirit, Flores and Oil out of Salt Nitre and Regulus Martis.

Take one part of REGULUS MARTIS STELLATUS (made of one part Iron or Steel, and three parts of Antimony, whose preparation is described in the fourth part) and three parts of pure salt nitre, mix and grind all together, and cast it in by little and little to distil, and there will come over a spirit together with a white sublimate, which must be separated with water, as has been taught above with other flores, and both the spirit and the flores are good to provoke sweat. The remaining CAPUT MORTUUM, (as they usually call it) is not dead, but full of life and virtue, whereby much good may be done both in Physick and Alchemy, as follows. The remaining Mass, which looks white, and is very sharp, and fiery (if the REGULUS was pure, if not, then it will look yellowish) may be edulcorated with fresh water, and it will yield a LIXIVIUM or lye in all things like unto calcined Tartar, but sharper and purer, and may be used almost in all operations instead of salt of Tartar (but first the REGULUS ANTIMONIS must be precipitated from it by the help of water) and afterward it may be coagulated into salt and kept for its use; the edulcorated, as also that which was precipitated with water is a white and fine powder, useful in the plague, fevers, and other diseases to provoke sweating thereby, and may very safely be used, and although if it be given in a greater quantity than usual, it causes some vomits also, yet for all that it does no hurt. It is easily taken because it has no taste. It is given to children from 3, 4, to 12 grains: to elder folks from 1/48 to 1/16 of an ounce; they work successfully in all diseases, where sweating is needful. This ANTIMONIUM

DIAPHORITICUM, may also be melted into glass, and so extracted and dissolved with spirit of salt, and it may be prepared into several good medicaments: and if all that which may be done with it, should be described at large, it would require too much time. The LIXIVIUM, if it be coagulated, has wonderful virtues, so that if one should describe them, he would hardly be credited by anybody, because it is not made of costly things; and truly the life of man is too short to find out by experience all that lies hid in it: and it would be but a laughing matter to a proud fool, if one should reveal it: therefore, it is better to keep counsel, than to sow strife. BASILIUS VALENTINUS in his **Triumphal Chariot of Antimony**[12], where he writes of the signed star, hinted it sufficiently, but very few take notice of it. PARACELSUS also, here and there in his books under an unknown name, makes frequent mention of it; but its true preparation and use, because of the unthankful was never described by the Philosophers, which for instruction of Good Honest Men we do here mention.

Before you edulcorate the REGULUS (made by fulmination) you may extract of it a good medicinal Tincture with spirit of wine, and if you dissolve it with spirit of salt, there will shoot a white foliated Talck in all things like unto the Mineral Talck: wherefore a liquor may be made, which colors the skin very white, but if this calx of Antimony, before it be extracted with spirit of wine or dissolved with spirit of salt be made into fine powder, and exposed to the moist air, it will dissolve into a fat liquor, which though it be something sharp, yet does no hurt to the skin, if it be used with discretion, but rather cleanses it more

---

[12] Vol. 2, R.A.M.S. Library of Alchemy.

than any other thing, and so it does likewise to the hair and nails; but as soon as the liquor has been applied for that purpose, it must be washed off again with water, lest it do not only take away the gross and unclean skin, but also work upon the tender white skin and do hurt, and therefore I give warning, that you use it discreetly: for according to the old proverb, you may misuse even that which else is good in itself. If you put some of it into warm water and bathe yourself in it, the gross skin will peel off the body, so that you will almost seem to be another body. And this bath also is good for many diseases: for it opens the pores mightily, and cleanses all the blood in the body, by drawing many ill humours out of it, which makes a man light and strong, especially if he be purged first, before he uses the bath. It is also good for Melancholy, scurvy and leprosie, especially when the red Tincture drawn out of it with spirit of wine, be used besides. It is also good to be used in a foot bath for those that are troubled with corns and other excrescencies upon their feet, or with nails that cut the flesh; for it softens them and makes them fit for cutting, and as tractable as wax. For there is nothing known under the Sun, which softens more a hard skin, hair, nails and other excrescencies, than this oil. And this I did set down therefore, because I know, that many are so tormented therewith, that they cannot well endure their shoes upon their feet. But if you coagulate this oil into salt, and melt it in a crucible, and pour it out into a flat brass basin, that it flows at large and may be broken, then you have the best Causticum, to open the skin withal where is need. If you dissolve crude Tartar with it and coagulate it again, you will get a salt which is used in many

Chymical operations; and there may be extracted out of it a blood red Tincture with spirit of wine, which proves very effectual against all obstructions.

    Also, every combustible sulphur may be easily dissolved with it, and used among bathes, it acts his part admirably. If any oil of spices be boiled therewith, then the oil will dissolve in it, and they turn together to a balsome, which does mingle itself with water, and is good to be taken inwardly for some infirmities: but women with child must not meddle with it, because it makes them miscarry. But after their delivery, it is good to expel after burthen and other reliques. But if you boil OLEUM with this liquor and rose water so long till the oil do incorporate with the liquor and waters and then separate the watery substance from it, you will get a soap as white as snow, which may be used for to wash the hands with it, and it does smell very well. You may also wash the head with it; for it strengthens the brain and cleanses the head and hair. This soap may be distilled, and it will yield a penetrating oil, very good for the sinews and nerves.

    Now as this liquor of REGULUS ANTIMONII softens the skin, nails, hair, feathers, horns, and the like, and dissolves them more than anything in the world: In the like manner, also it has power to dissolve not only metals, but also the hardest stones, but not in that manner which is done by boiling, as was mentioned of sulphur, but after another way, which is not proper for this place. It suffices that I hinted it.

    The fiery fixed salt nitre may be dissolved with spirit of salt or vinegar, and sublimed into a TERRA FOLIATA. What further can be effected with it, does not belong to this place, and perchance somewhere else more shall be spoken of it.

To Distil a Butrum out of Antimony, Salt and Vitriol, like unto that, which is made from Antimony and Mercury Sublimate.

Take one part of crude Antimony, two parts of common salt, and four parts of vitriol calcined white, beat all to powder and mix them well, and so cast it in as you were taught to do with other materials, and there will come over a thick oil of Antimony like butter, which may be rectified like any other oil, that is made after the common way with Mercury sublimate, and is also the same with it in use, which use you may see in the first part: the same also may be made better and in a greater quantity in the furnace described in the first part, and also with less coals and time by the help of the open fire, because it endures greater heat than in the second furnace.

To Distil Butrum of Arsenick and Orpiment.

After the same manner, as was taught with Antimony, there may also out of Arsenick and Auripigment together with salt and vitriol a thick oil be distilled, which not only outwardly but also inwardly is safe to be used, and may be so corrected, that it shall be nothing at all inferior in virtue unto the BUTYRUM ANTIMONII, but rather go beyond it: which perchance will seem impossible to many. But he that knows the nature and condition of minerals, will not be astonished at my words, but they will be to him as a light in a dark place.

To Make a Rare Spirit of Vitriol.

    If common vitriol be dissolved in water, and you boil granulated Zinck in it, all the metal and sulphur contained in the vitriol will precipitate on the Zinck, and the solution will turn white, the precipitated matter is nothing else, but iron, copper, and sulphur, which the salt of vitriol did contain, and now is drawn from it by the Zinck.

    The reason why the metal precipitates out of the salt upon the Zinck, belongs to the fourth part, where you will find it sufficiently explained; The white solution, from which the metallical matter is separated, must be coagulated to the dryness of salt, and so by itself a spirit distilled of it, which rises easily, and is in taste and virtue not unlike unto common oil of vitriol, but only that this is a little purer than the common.

    Here perchance many may object: you take the green from the vitriol, which PARACELSUS does not teach, but bids us to keep it. To which I answer, that I do not teach here to make the sweet red oil of vitriol, whereof PARACELSUS has written, but the white acid oil; which is as good, or rather much better, than common oil, which is made of the common impure vitriol. To what purpose is it, that you take green vitriol to distil, whereas the green does not come over, and although that green should come over, why should that oil be better than the white? For the green in the common vitriol is nothing else but copper and iron, which the salt water running through the passages of Metals did dissolve and take into itself, and as soon as such a green vitriol feels the fire, the green turns into red, which is nothing else but a calcined iron or copper, which in the reducing by a strong fire and by melting is made manifest.

PARACELSUS has not taught us, that we should drive over the green by the force of the fire into a red and sweet oil, but he has showed us another way, which is found out by few men, whereof in the beginning of the second part already has been made mention.

This spirit or acid oil distilled out of the purified vitriol, is of a pleasant sourness, and serves for all those uses, which above by the vitriol were described. And this process is set down only for that end, that we may see, that when the vitriol is purified, that then it is easier distilled, and yields a more pleasant spirit, than if it be yet crude and impure.

And that such a purifying of the vitriol is nothing else but a precipitating of the metal, which the water (as before said) running through the veins thereof has assumed, is thus to be proved; dissolve any metal in its appropriate Menstruum, whether it be done with distilled acid spirits or sharp salts, adding common water to them, or else dry by the fire in a crucible, according as you please, and then put into that solution another metal, such as the dissolvent does sooner seize on, then upon that which it has assumed, and then you will find, that the dissolvent does let fall the assumed metal or mineral, and falls upon the other, which it does sooner seize on, and dissolves it as being more friendly to it; of which precipitation in the Fourth Part shall be spoken more at large.

This one thing more is worthy of your observation, that among all metals there is none more soluble than Zinck, and therefore that all the other (as well in the dry as in the wet way) may be precipitated thereby and reduced into light calxes, in so much as the calx of gold or silver precipitated in this manner (if so you proceed well) retains its splendor or gloss, and is like a fine powder wherewith you may write out of a pen.

To Make a Subtle Spirit and Pleasant Oil of Zinck.

Because I made mention here of Zinck, I thought good not to omit, that there may be made a penetrating spirit and wholesome oil out of it by the help of vinegar, which is thus to be done. Take of the flores (which were taught to be made in the first part) one part, put them into a glass (fit for digestion) and pour upon them 8, or 10 parts of good sharp vinegar made of honey; or in want thereof take wine vinegar, and set the glass with the flores and vinegar in a warm place to dissolve, and the solution being performed, pour off the clear, which will look yellow and after you have filtered it abstract the phlegm, and there will remain a red liquor or balsome, to which you must add pure sand, well calcined, and distill it, and first there will come over an unsavory phlegme, afterward a subtle spirit, and at last a yellow and red oil which are to be kept by themselves separated from the spirit, as a treasure for to heal all wounds very speedily. The spirit is not inferior unto the oil, not only for inward use to provoke sweat thereby, but also externally for the quenching of all inflammations, and doubtless this spirit and oil is good for more diseases but because its further use is not known to me yet, I will not write of it, but leave the further trial to others.

To Distil a Spirit and Oil out of Lead.

In the same manner, as was taught of Zinck, there may be out of lead also distilled a subtle spirit and a sweet oil, and it is done thus: Pour strong vinegar upon MINIUM[13], or any other calx of lead, which is made per se, and not with sulphur, let it digest and dissolve in sand or warm ashes, so long till the vinegar be colored yellow by the lead, and turned quite sweet. Then pour off the clear solution, and pour on other vinegar, and let this likewise dissolve, and this repeat so often. till the vinegar will dissolve no more, nor grow sweet; then take all these solutions, and evaporate all the moisture, and there will remain a thick sweet yellow liquor, like unto honey, if the vinegar was not distilled and made clear, then no liquor remains, but only a white sweet salt. This liquor or salt may be distilled after the same manner as was taught with the Zinck, and there will come over not only a penetrating subtle spirit, but also a yellow oil, which will not be much, but very effectual, in all the same uses, as of the spirit, and oil of Zinck was taught.

N. B. This is to be observed, that for to make this spirit and oil, you need no distilled spirit, but that it may be done as well with undistilled vinegar, and the undistilled yields more spirit than the distilled. But if you look for a white and clear salt, then the vinegar must be distilled, else it does not shoot into crystals, but remains a yellow liquor like unto honey, and it is also needless to make the solution in glasses, and by digestion continued for a long time, but it may as well be

---

[13] Minium is the naturally occurring form of lead tetroxide, $Pb^{2+}_2Pb^{4+}O_4$ also known as red lead. -PNW

done in a glazed pot, viz. pouring the vinegar upon the Minium in the pot, and boiling it on a coal fire; for you need not fear that anything of the vinegar will evaporate, in regard that the lead keeps all the spirits, and lets only go an unsavory phlegm. You must also continually stir the lead about with a wooden spatula, else it would turn to a hard stone, and would not dissolve: the same must be done also when the solution is done in glasses; and the solution after this way may be done in three or four hours: and when both kinds of solutions are done, there will be no difference betwixt them, and I think it providently done not to spend a whole day about that which may be done in a hour.

And if you will have this spirit and oil better and more effectual, you may mix 1 ounce of crude Tartar made into powder with 1 lb. of dissolved and purified lead, and so distil it after the same manner as you did distil it by itself, and you will get a much subtler spirit and a better oil than if it were made alone by itself.

To Distil a Subtle Spirit and Oil out of crude Tartar.

Many think it to be but a small matter to make the spirit of Tartar; for they suppose, that if they do but only put Tartar into a retort, and apply a receiver, and by a strong fire force over a water, they have obtained their desire: and they do not observe, that instead of a pleasant subtle spirit, they get but a stinking vinegar or phlegm; the pleasant spirit being gone. Which some careful operators perceiving, they caused great receivers to be made, supposing by that means to get the spirit. Now when the distillation was done, weighing their spirits together with the remainder, they found, that they had suffered great loss, wherefore they supposed it to be an impossible thing, to get all the spirits, and to lose none, and indeed it is hardly possible to be done otherwise by a retort: for although you apply a great receiver to a small retort, and that there be also but a little Tartar in it, and the joints being well luted, so that nothing can pass through, and though you make also the fire ever so gentle, hoping to get the spirit by that way, yet for all that you cannot avoid danger and loss. For at last the retort beginning to be red hot, and the black oil going, then and but then the subtlest spirits will come forth, which either steal through the joints, or else do break the retort or receiver, because they come in abundance and with great force, and do not settle easily: wherefore I will set down my way of making this most profitable, and excellent spirit.

The Preparation and Use of the Spirit of Tartar.

Take good and pure crude Tartar, whether it be red or white, it matters not, make it into fine powder, and when the distilling vessel is red hot, then cast in with a ladle half an ounce and no more at once, and so soon as the spirits are gone forth and settled, cast in another half-ounce and this continue, till you have spirit enough, then take out the remainder, which will look black, and calcine it well in a crucible, and put it in a glass retort, and pour the spirit that came over together with the black oil, upon it, drive it in sand at first gently, and the subtlest spirits will come over, and after them phlegme, at last a sour vinegar together with the oil, whereof you must get each by itself. But if you desire to have the subtle spirit which came over first, more penetrating yet, then you must take the CAPUT MORTUUM that stayed in the retort, and make it red hot in a crucible, and abstract the spirit once more from it, and the calcined Tartar will keep the remaining moistness or phlegm, and only the subtlest spirit will come over, which is of a most penetrating quality, whereof from half a dram to an ounce taken in wine or any other liquor provokes a quick and strong sweat, and it is a powerful medicine in all obstructions, and most approved and often tried in the plague, malignant fevers, scurvy, MELANCHOLIA HYPOCANDRIACA, collick, contracture, epilepsy and the like diseases. And not only these mentioned diseases, but also many others more, which proceed from corrupt blood under God may successfully be cured with it.

The phlegm is to be cast away as unprofitable: the vinegar cleanses wounds: the oil allays swelling and pains, and does cure scabs, and disperses knobs that are risen upon the skin, as also other excrescencies of the same, if it be used timely, and the use thereof be continued.

N. B. If the black stinking oil be rectified from the calcined CAPUT MORTUUM, it will be clear and subtle, and it will not only assuage very speedily all pains of the gout, but also dissolve and expel the conglobated gravel in the reins, applied as a plaister or unguent. In like manner, it will dissolve and extract the coagulated Tartar in the hands, knees and feet, so that the place affected will be freed and made whole thereby: because in such a despicable oil there lies hidden a volatile salt which is of great virtue. But if you desire experimentally to know whether it be so, then pour upon this black stinking oil an acid spirit, as the spirit of common salt, or of vitriol or salt nitre, or only distilled vinegar, and the oil will grow warm and make a noise and rise, as if AQUA FORTIS had been poured upon salt of Tartar, and the acid spirit will be mortified thereby, and turn to salt. And this well purified oil does dissolve and extract the Tartar out of the joints (unless it be grown to a hard-stony substance) even as soap scours the uncleanness out of cloths, or to compare it better, even as like receives its like, and is easily mixed with it; and does love it; but on the contrary, nothing will mix itself with that wherewith it has no affinity at all. As if you would take pitch out of cloth by washing it in water, which never will be done because of the contrary nature; for common water has no affinity with pitch or other fat things, nor will it ever be taken out therewith without a mediator, partaking of both natures, viz. of the nature of pitch and that of the water, and such are sulphureous salts, and nitrous salts, whether they be fixed or volatile. As you may see at the soap boilers, who incorporate common water by the help of sulphureous salts with fat things, as tallow and oil. But if you take warm oil or any thin fat substance, and put it upon the pitch or rozin, then the oil easily accepts of and lays hold on its like, and so the pitch is dissolved and gone out of the cloth, and the remaining fatness of the oil may be fetched out of the cloth with lye and common water or soap, and so the cloth recovers its former beauty and pureness. And as it falls out with sulphureous things, so it does likewise with Mercurial. For example, if you would take the salt out of powdered flesh or pickled fish with a lixivium it would not succeed, because that the nitrous and acid salts are of contrary natures.

But if upon the powdered flesh or pickled fish you pour on water wherein some of the same salt (wherewithal the flesh was powdered) is dissolved, that salt water will extract the salt out of the flesh, as being its like, much more than common water, wherein there is no salt.

In this manner, the hardest things also, as stones and metals, may be joined or united with water, whereof more in my other books are extant; it is needless here therefore to relate. I gave a hint of it, only for to show, that always like with like must be extracted. True it is that one Contrary can mortifie another, and take the corrosiveness from it, whereby the pains for a time are assuaged, but whether the cause of the disease itself be eradicated thereby, is a question.

Here may be objected, that I make a difference between the sulphureous and Mercurial salts, whereas neither Mercury nor sulphur apparently is to be seen in either. It is true, he that does not understand nor know the nature of salts, is not able to apprehend it. And I have not time now to demonstrate it, but the same is showed at large in my book DE NATURA SALIUM, that some of them are sulphureous, and some Mercurial: but he that looks for a further direction yet, let him read my book DE SYMPATHIA & ANTIPATHIA VERUM, wherein he shall find it demonstrated that from the Creation of the World to the time present, there were always two contrary natures fighting one against the other, which fight will continue so long till the Mediator betwixt God and Man, the Lord Jesus Christ shall put an end unto this strife, when he shall come to separate the good from the bad, by whose lightning and fire flame the proud and hurtful superfluous sulphur shall be kindled and consumed: the pure Mercurial being left in the center.

How to Make Precious Spirits and Oils out of Tartar joined with Minerals and Metals.

Take any metal or mineral, dissolve it in a fit menstruum, mix it with a due proportion of crude Tartar, so that the crude Tartar being made into powder together with the solution make up a pap as it were; then at once cast in one spoonful of it, and distil it into a spirit and oil, which after the distillation must be separated by rectification, for to keep each by itself for its proper use.

The Use of the Metallized Spirit and Oil of Tartar.

This Tartarized spirit of metals is of such a condition, that it readily performs its operation according to the strength of the spirit, and the nature of the metal or mineral, whereof it is made. For the spirit and oil of Gold and Tartar is good for to corroborate the heart, and to keep out its enemies: the spirit of silver and tartar does serve for the brain; that of Mercury and Tartar, for the liver: of lead and tin, for the spleen and lungs: of iron and copper for the reins and seminary vessels: that of antimony and tartar for all accidents and infirmities of the whole body; and these metallical spirits made with Tartar, provoke sweat exceedingly, whereby many malignities are expelled out of the body. Likewise, also the oil has its operation, though this of several metals, as of Mercury and Copper, is not well to be used inwardly, because it causes salivations and strong vomits. But externally they are very good to cleanse all putrid ulcers, and to lay a good and firm ground for healing them.

The remainder, whereof the spirit and oil is distilled, you may take out, and reduce it in a crucible into a metal, so that what is not come over, may not be lost, but made to serve again.

And as you were taught to distil spirits and oils out of dissolved metals and crude Tartar; so you may get them likewise out of common vitriol and Tartar, viz. thus, take one part of Tartar made into powder, two parts of good pure vitriol, mix them well together, and distil a spirit of them, which though it be unpleasant to take, for all that in all obstructions and corruption of blood whatsoever it is not to be despised, but very successfully performs its operation; especially when it is rectified from its CAPUT MORTUUM, and so freed from its phlegm; and its best virtue, which consists in the volatility, be not lost in the distilling.

N. B. But if you will have this spirit more effectual, then you may join Tartar and vitriol by boiling them together in common water, and crystallizing; and then cast it in, and distil it, and there will come over a much purer and more penetrating spirit; because that is the solution and coagulation of both, many feces were separated: but if one part of vitriol you take two parts of Tartar, and dissolve it together, and so filter and coagulate it, then the Tartar with the vitriol will shoot no more, but there remains a thick liquor like unto honey, out of which with spirit of wine there may be extracted a good tincture against obstructions. This liquor taken from $1/24$ to $1/8$ ounce does purge very gently, and sometimes it causes a vomit, especially if the vitriol was not pure and good: and it may be also distilled into a spirit not inferior unto the former in virtue. Besides the way taught above, there is yet (for to distil a metallized spirit of Tartar) another way, whereby several metals and minerals may be reduced into pleasant spirits and oils, and of more virtue, and it is done in this manner.

Take of the Tartar of white Rhenish wine made into powder, pour upon it sweet rain or running water, so that to 1 lb. of tartar there be 10 lb. or 12 lb. of water, or so much that the tartar may be dissolved by it in the boiling, and then boil the mixture with the water in a tinned kettle, or which is better, in a glazed pot, until it be quite dissolved, and in the mean while take off the scum (with a wooden skimmer) still as it rises in the boiling: and when no more scum rises, and all the tartar is dissolved, then pour the solution thus hot through a linen cloth, tied straight on an earthen glazed vessel, that the remaining sliminess may be separated. The tartar water being strained, let it stand for 24 or 30 hours without stirring, and there will stick a crystallized tartar to the sides of the vessel, which after the water is poured off may be taken out, and washed with cold water, and then dried. This purified tartar keep, until I shall teach you what further is to be done with it; and this tartar is pure enough for the above said purpose, viz. to reduce metals into oil with it, as shall follow anon. It is also good taken of itself for an abstersive to make the body soluble. But if you desire to have it yet whiter and fairer and in great Crystals, you must proceed thus.

You must know this that all salts, if they shall shoot into great crystals, there must be a great quantity of them, for of little there comes but little. And if you will make great and fair white crystals of tartar, which will be no better than the former, but only pleasant to the eye, then you must proceed in this manner.

Take of white tartar made into powder, about ten or thirty lb. Pour so much water upon it, as is needful for to dissolve it, and boil it by a strong fire in a tinned kettle, until all the tartar be dissolved, which you may know by stirring it with a wooden ladle, and skim off diligently all the filth rising on the water; and you must take heed, that you take neither too much nor too little water to it; for if you take too little, part of the tartar will remain undissolved, and so will be cast away and lost among the slime: but if you take too much of it, then the tartar is too much dispersed in the water, and cannot shoot well, and so will likewise be lost, being cast away afterwards with the water, For I have heard many a one complain, that they could get but little of a pound, and therefore supposed the tartar to have been naught, whereas the fault was not in the tartar, but in the workman, that managed not well his work, pouring away one half which did not shoot with the water: but if you proceed well, then four pounds of common tartar will yield 3 lb. of pure white crystals. The solution being well made, and no scum more rising at the top, cover the kettle, and let it cool without removing from the warm place it stands in, which will be done within three or four days, if the kettle be big. But the fire must be taken away from under the kettle, and so let it stand for the time mentioned. In the meantime while the Tartar will crystallize to the sides of the kettle, which crystals after the time is expired, and the water poured off, are to be taken out and washed and boiled again with fresh water, and so skimmed and crystallized; and this proceeding must be still reiterated, until (which is done the third or fourth time) the crystals are white enough: then take them out, dry and keep them for use; whereof from 1/8 to 1 ounce made into powder, and taken in wine, beer, warm broth or other liquor, will give some gentle stooles, and serves for those, which cannot endure strong-physick. This tartar may be sharpened with Diagridium or any other purging drug, that so you need not take it in so great a quantity at once, but a lesser dose may serve turn. But if you do not look for great crystals, but only for Tartar well purified, then you may use this following manual, and you will get exceeding fair and glistering crystals, which need no beating into powder, but by the working come to be so pure and fine, as if they had been ground upon a stone, and looking not like a dead powder, but having a gloss, like unto small glistening snow that fell in very cold weather, and it is done thus: when the crystals are come to be pure enough by often dissolving and coagulating, then dissolve them once again in pure water, and pour the solution into a clean vessel of wood, copper, or earth being glazed; and let it not stand still (as above taught with the crystals) but as soon as it is poured in, with a clean wooden stick stir about continually without ceasing, till all be cold, which will be done in half an hour. In this stirring the Tartar has no time to shoot into crystals, but does coagulate into the smallest glistening powder, pleasant to behold, and like unto frozen snow settles at the bottom of the vessel, then pour off the water, and dry the powder, and keep it for use. The waters which you poured off, in regard that they contain yet some Tartar, ought not to be cast away (as others do) but evaporated, and the Tartar contained in them will be saved, and so nothing will be lost, and in this manner not only white Tartar may be reduced into clear crystals, but also the red being several times dissolved and crystallized, loses its redness, and turns white and clear. Besides the above said, there is another way to reduce the Tartar into great white crystals at once by precipitation; but these being good enough for our purpose, viz. to make good medicines out of metals, I hold it needless to lose more time by the relation of it, and so I will acquiesce.

Another Way to Make a Metallized Spirit of Tartar.

Take of purified Tartar dissolved and coagulated but once, as much as you please, pour so much rain or other sweet water to it as will serve to dissolve it; in which solution you must boil plates of metals, until the Tartar has dissolved enough of it, so that it will dissolve no more; the sign whereof is, when the solution is deep colored of the metal, and during your boiling you must often supply the evaporated water with pouring on of other, lest the Tartar come to be too dry and burn; and this solution may be done best of all in a metallical vessel; as when you will make the solution of iron, you may do it in an iron pot; and for copper you may take a copper kettle, and so forth for other metals, a vessel made of the same is to be taken. But you must know that gold, silver, and crude Mercury, unless they be first prepared cannot be dissolved like iron and copper, but when they are prepared first for the purpose, then they will also be dissolved. In like manner, some minerals also must be first prepared before they can be dissolved with Tartar and water. But if you can have good glasses or glazed vessels of earth, you may use them for all metals and minerals for to dissolve them therein, and the solution you may not only use of itself for a medicine, but also distil it, and make a very effectual spirit and oil of it as follows.

To Distil the Spirit and Oil of Lead and Tin.

Take the filings of Lead and Tin, and boil them with the water or solution of Tartar in a leaden or tin vessel, until the Tartar be sweetened by the water, so that it will dissolve no more, to which pass it will be brought within twenty four hours, for both these metals will be dissolved but slowly, but if you would perform this solution sooner, then you must reduce the metals first into a soluble calx, and then they may be dissolved in less time than an hour. The solution being done, you must filter it, and in B. abstract all the moisture to the thickness or consistency of honey, and there will remain a pleasant sweet liquor, which of itself without further preparation may safely be used inwardly for all such diseases, for which other medicaments, made of these metals are useful. Especially the sweet liquor of lead and tin does much good in the Plague, not only by driving the poison from the heart by sweating, but also by breaking or allaying the intolerable heat, so that a happy cure does follow upon it: but externally the liquor of lead may be used successfully in all inflammations, and it heals very suddenly, not only fresh wounds, but also old ulcers turned to fistulas; for the Tartar cleanses, and lead consolidates.

The liquor of tin is better for inward use than for outward whose operations is not so fully known yet, as that of lead. But if you will distil a spirit thereof, then cast it in with a ladle by little and little, as above in other distillations oftentimes was mentioned, and there will come over a subtle spirit of tartar, carrying along the virtue and best essence of the metal, and therefore does also prove much more effectual than the common spirit of tartar, which is made alone by itself, and this spirit as well that which is made of tin, as that of lead, if it be well dephlegmed first, may be used and held for a great treasure in all obstructions, especially of the Spleen; and few other medicines will go beyond them; but besides there must not be neglected the use of good purging medicines, if need require them. With the spirit there comes over also an oil, which is of a quick operation, especially in wounds and sores of the eye, where other ointments and plasters may not so fitly be used, for it does not only allay the heat and inflammation, a common symptom of the eye wounds, but also does hinder and keep back all other symptoms which few other medicaments are able to do; and for the residue, if it be driven further by the strongest fire, then there will come over a sublimate, which in the air dissolves into oil, which is also of a powerful operation, not only in physick, but also in Alchemy.

And the lead runs together into a fair white REGULUS, which is much whiter, purer and fairer than other common lead: but the tartar retains the blackness, and raises itself to the top as a fusible dross, which is impregnated with the sulphur of lead, wherewith you may color hair, bones, feathers and the like, and make them to be, and remain brown and black.

    I made trial once of such a distillation in an iron vessel, whereby the same in the inside was so whitened by the purified lead, that it was like unto fine silver in brightness; which afterwards trying again, it would not fall so fair as at first; whereat none ought to wonder, for I could write something more (if it were fit) of tartar, knowing well what may be effected with it, if I did not stand in fear of scoffers, which vilify all what they do not understand. I durst presume to call tartar the Sope of the Philosophers; for in the cleansing of some metals, by long experience I found it of admirable virtue; though I would not be understood thus, as if I did count it to be the true AZOTH UNIVERSALIS PHILOSOPHORUM, whereby they wash their Laton: but I cannot deny, but that it is of particular use for the washing and cleansing of several metals; for it is endued with admirable virtues for the use of metals, whereof in other places more shall be said hereafter.

How to Make a Tartarised Spirit and Oil out of Iron or Steel and Copper.

    If you intend to make a good medicine out of iron or steel or copper joined with tartar, then for the iron or steel take an iron pot, and for copper a kettle of copper, make them very clean and put in it the filings of iron, or steel, or copper, which you please, and twice as much of pure tartar made into powder, and so much water, that the tartar may be dissolved well by it in the boiling, and so boil the metal with the tartar—water so long, till it be deeply colored by the metal, as red by the iron, and deep green by the copper; and when the water in the boiling does waste, you must still supply it with other, that the tartar may not burn; for there must be always so much water, that no skin of the tartar may rise at the top, but that it remain always open, and there must not be too much water neither, lest it be too sweet, and not able, to dissolve the metal. The solution of iron or steel being come to be red and sweet, and in taste like unto vitriol, but green and bitter of copper, pour it off warm by inclination into another clean vessel, and let it stand so long again in a very gentle heat of coals, till almost all the water be evaporated, and the dissolved metal with the tartar remain in the consistency of honey.

Which metallical liquor may be used inwardly and outwardly (especially that of iron) which does purge gently, and opens the obstructions of the Liver and Spleen: cleanses the Stomach, and kills Worms: externally used it is a good wound balsome, and goes far beyond all such as are made of vegetables. It is a singular treasure, not only for to cure new wounds; but also for to cleanse and heal old corrupt exulcerated sores, turned to fistulas; but the liquor of copper is not safe for to be used inwardly, for it is not only very unpleasant in taste, but also causes vehement vomits: and therefore I would not advise any one to be forward to use it, unless it be for strong folks and for to kill worms in them, for which purpose it is excellent and surpasses all other medicines whatsoever; but to little children it ought not to be given at all, in regard that it is of far too strong an operation for them.

N. B. And if you will use it to strong bodies against the worms or stomach-agues, you must observe that the patient (in case that he cannot get it up) thrust his finger into the throat to further the vomiting, that it may not stay behind, but come forth again out of the body, which if done, health follows upon it; but if it remains in the body, it causes a loathsomeness to use it any more. And therefore you must take heed to use it warily: and in regard that this liquor is very bitter, you may mix it with some sugar, to facilitate the taking thereof, but that of iron needs no such correction, it being sweet enough of itself, and therefore I commend and prefer it before the other: but if you will needs have that of copper (because it works so strongly) then the Patient must keep in from the cold air, and not presently after the operation load the stomach with strong drink and superfluity of meat, contenting himself with some warm broth and a little cup of wine or beer, and the next day his meat and drink will taste the better with him, and do him so much the more good.

But externally, this liquor is of the same use with that of iron or steel, yea, proves more effectual and speedier in healing. It would be good that Surgeons knew how to prepare it, and would use it instead of their salves, wherewith many fresh wounds are spoiled and turned into horrid ulcers, especially it requiring so little cost and pains to make it. And if you would have these liquors purer yet, you must pour spirit of wine, and extract them, and they will easily yield their tincture, and leave many faeces behind which are good for nothing: but the tincture will be so much the better, purer, and more effectual, so that you need use but four or five drops for purging, whereas of the gross liquor you must have from 4, 6, 8, to 12 or 16 drops: and this extracted tincture works also much better externally, and keeps longer than the balsome or liquor; which in time is corrupted, but the extraction is never spoiled. But if you will distil the liquor or balsome, it is needless that it be extracted first, but may be distilled as the boiling made it, after the same manner, as above was taught for Lead, and there will come over a yellow spirit and oil from iron or steel, and from copper a greenish spirit and oil.

The spirit and oil of iron may safely be used in the plague, fevers, obstructions, and corruptions of the blood, from 1/8 to 1 ounce. It is much better to provoke sweat, than that which is made of crude Tartar, without addition of a metal: the like does that also which is made of copper and more effectually yet, and sometimes causes a vomit, if it be used in a greater quantity, than is fitting.

N. B. Although the Chymists do prefer copper before iron, as a more firm and ripe metal, nevertheless, it is found by experience, that iron or steel because of its sweetness is better to be used for an inward Medicine than copper. But for external use, copper (if it be well prepared) has the pre-eminence, being an appropriate medicine for all ulcers and open sores, in all parts of the body, if the same inwardly be kept clean by fitting purges. For not only the now described medicine, but also many more besides, are taught to be made from copper in other places of my books.

A Country—physick and purge I will teach for those, which either live far from Apothecary-shops, or have no money to spare for physick; and it is made out of iron and copper, whereby they may cleanse their slimy stomachs, spoiled by a disorderly diet, whence head—aches, worms, agues, and other diseases are occasioned, warning withal those that are either too old or too young, or else decayed and weak, and so not strong enough for such powerful physick, that they will forbear to use it, lest besides the worms, they kill and expel life itself also; but those that are of a strong constitution, and a middle age, and of a sound heart, may safely use this purge, whereby stomach-agues, belly-worms, and many other occult diseases may be cured with good success. The preparation is done thus: Take 1 ounce of pure tartar made into powder, & 1 ounce or 1/2 ounce of sugar or honey, and 5 ounces or 6 ounces of spring water or rain, water, put all into a clean copper vessel which is not greasy, and boil it upon a coal fire as long or somewhat longer than you use to boil an egg, or at the furthest half a quarter of an hour; take off the scum in boiling, let it stand till it be milk-warm, so that it may be drunk. This potion tasting almost like warm wine sweetened with sugar, give unto the patient to drink, and let him fast upon it, and within half an hour it will begin to work upwards and downwards; whereat you need not be amazed, but only keep the body warm, and within an hour it will have done working. But if you will drive out worms from little children by purging, then instead of the copper-vessel, take a clean iron-vessel, and put in a less quantity of tartar, sugar and water, and boil it as above said, and give it to them, and it will purge only downwards, but sometimes it will also give a gentle vomit, which will do them no hurt, but rather will cleanse the stomach the better. But if the drink be too weak, so that it does not work, it may be used again the next day (but you must take more of the ingredients, or else let them boil longer). There is no danger in it at all, if you proceed aright, and it is much pleasanter to take, than the bitter worm-seed, wherewith they usually torment children.

The reason why this decoction works in this manner is, that the tartar and sugar being boiled in metallical vessels with water, work upon the metal, and extract virtue out of it, which causes vomiting and purging (the Tartar also being helpful to it).

How to Make a Tartarised Spirit of Mercury.

Vulgar Mercury cannot be dissolved like the former metals with tartar and water, without any foregoing preparation; but must be sublimed first with salt and vitriol, or crystallized with AQUA FORTIS, and then it may be dissolved by boiling with tartar and water, and reduced into a balsome, like other metals, but is not to be used inwardly, unless it be digested a sufficient time, so that its fierceness be allayed. Externally it may safely be used in all desperate, especially venereal sores, and it is very effectual and profitable medicine for them. But most of all it does serve for Alchemy, although few do know this guest, because he will not be seen by everyone. The spirit which comes over from it by distillation, is an admirable thing not only in physick, but also in Alchemy: yet you must take heed, that instead of a friend, you do not harbor a great enemy; for its force and virtue is very great and powerful.

How to Make a Tartarised Spirit of Gold and Silver.

Gold and silver also can by no means be dissolved with tartar in a wet way: but in a dry way adding its helper to it, it will easily dissolve, which does not belong hither; but if you will draw a spirit of it, then the gold and silver must first by dissolving and coagulating be reduced to crystals, and dissolved with purified tartar and water, and of Gold you will get a yellow solution, and of silver a white inclining unto green, which being reduced to the consistency of honey, may be used safely and without fear. The solution of Gold does loosen and keep the body open; it effectually strengthens the stomach, heart, lungs, and liver, and other principal members: and that of silver purges very forcibly, according to the quantity given, like another purge, but without harm or danger, so that in all diseases where purging is necessary, it may be used safely from 1/24 to 1/8 ounce but that of gold is used in a smaller quantity: and both the liquor of gold and of silver may very successfully be used externally: but because for external uses inferior metals will serve the turn, it is needless to use costly things thereto.

    The spirit which is forced from it by distillation, is endued with great virtue: for the volatile part of the metal comes over joined with the spirit of tartar, the remainder may be reduced, so as it was taught of other metals. This spirit, especially that of Gold, is exceedingly good in the plague and other diseases, where sweating is necessary: for it drives not only by sweating, all Malignities from the Heart, but also does strengthen the same, and preserves it from all hurtful symptoms. Likewise, also that of silver is very commendable, especially if it be first dephlegmed from its CAPUT MORTUUM, as above was taught in the preparation of the common spirit of tartar. For any Physician expert in Chymistry may easily guess what the spirit of tartar well rectified and impregnated with the virtues of gold may effect, and therefore it is needless to make any further mention of it, but it shall be left to the trial thereof.

To Make a Tartarised Spirit of Antimony.

Crude Antimony cannot be dissolved in such a manner as above has been taught: but if it be first prepared into flores, or a VITRUM, it yields easily its virtue in boiling, and it is done thus: Take to one part of the flores or of small ground VITRUM ANTIMONII made PER SE, three parts of pure tartar, and 12, or 15 parts of clean water, boil the Antimony with the tartar and water in a glazed pot for three or four hours, and the evaporated water must be still supplied with other that the tartar may not burn for want of water, and the VITRUM must be sometimes stirred about with a wooden spatula (which the flores being light do not need). Thus done, the tartar water will be deep colored by the Antimony, and leave the remaining Antimony settled in the bottom, from which pour off the solution, and after having filtered it, evaporate the water from it, and then extract it once more with spirit of wine, and you will get a blood red EXTRACTION, whereof 1, 2, 3 to 10 or 12 drops given at once, causes gentle vomits, and stools, which may be safely used by old and young in all diseases that have need of purging, and you need not fear any danger at all. For I know of no other vomit-inducer, which purges more gently than this, and if you please, you may make it work only (PER INFERSORIA) downward, so that it shall cause no vomits at all: and you need do nothing else but make a toast of brown bread, and hold it hot to your nose and mouth, and when this is almost cold, have another hot in readiness, and so use one after another by turns, till you feel no more loathing, and that the virtue of Antimony has begun to work downward. This is a good secret for those that would use Antimonial physick, but that they are afraid of vomiting, which they are not able to endure. But if you will not spend so much pains, as to make such an Extract, then do as you was taught above to do with the copper, and take ten or twelve grains of prepared Antimony for an old body, but for a young one 5, 6 grains or more or less according to the condition of the person, and 1 ounce or 1/8 ounce of pure tartar, and together with 4 ounces of water put it in a little pipkin, and boil it a quarter of an hour, then pour the solution only into a cup, and dissolve a little sugar in it, whereby the acidity of the Tartar will be somewhat qualified. The DECOCTION drink warm, and keep yourself as it is fit, and it will work much better, than if it had been steeped overnight in wine, which not everyone can abide to take fasting; but this DECOCTION, because it tastes like warm and sweet wine, is much pleasanter to take.

    N. B. It is to be admired, that well-prepared Antimony is never taken in vain: for although it be given in a very small quantity, so that it cannot cause either stools or vomits, yet it works insensibly, viz. it cleanses the blood, and expels malignities by sweat, so that mighty diseases may be rooted out thereby without any great sensible operation. Which many times happened unto me, and gave me occasion to think further of it; and therefore I sought how to prepare Antimony so, that it might be used daily without causing of vomits or stools, which I put in execution accordingly, and found it good, as afterward shall follow.

Of the solution above described, viz. of the flores of Antimony with tartar make a good quantity, and after the evaporation of the water distil a spirit of it, and there will also come over a black oil, which must be separated from the spirit, and rectified PER SE, and externally applied it will not only do the same wonderful operations, which above have been ascribed to the simple oil of tartar, but it goes also far beyond it, for the best essence of Antimony has joined itself thereunto in the distilling and so doubled the virtue of the oil of Tartar, and this oil may with credit be used not only for all podagrical tumors to allay them very readily, but also by reason of its dryness it does consume all other tumors in the whole body, whether they be caused by wind or water: for the volatile salt by reason of its subtlety, conveys the virtue of Antimony into the innermost parts of the body in a marvelous and incredible way, whereby much good can be performed in Chirurgery.

As for the spirit, you may not only use it very successfully, in the Plague, Pox, Scurvy, MELANCHOLIA HYPOCONDRIACA, Fevers, and other obstructions and corruptions of blood, but also if you put some of it into new wine or beer, and let it work with it, the wine or beer comes to be so virtuous thereby, that if it be daily used, it does stay and keep off all diseases proceeding from superfluous humours and corrupted blood, so that neither Plague, Scurvy, MELANCHOLIA HYPOCHONDRIACA, or any other disease of that kind can take root in those that daily use it, wherein no metal or mineral (except gold) can be paralleled with it: but in case you have no conveniency to make that spirit, and yet you would willingly have such a medicinal drink made of Antimony, then take but of the solution made with tartar, before it be distilled, and put 1 lb. or 1/2 lb. of it into 18 or 20 gallons of new wine or beer, and let it work together, and the virtue of the Antimony by the fermentation of the wine will grow the more volatile and efficacious to work. And if you cannot have new wine (in regard that it does not grow everywhere) you may make an artificial wine of Honey, Sugar, Pears, Figs, Cherries or the like fruit, as in the following third part shall be taught, which may stand instead of natural Wine.

These medicinal wines serve for a sure and safe preservative, not only to prevent many diseases, but also if they have possessed the body already, effectually to oppose and expel them. Also, all external open sores (which by daubing and plastering could not be remedied) by daily drinking thereof may be perfectly cured. For not only BASILIUS VALENTINUS, and THEOPHRASTUS PARACELSUS, but many more before and after them knew it very well, and have written many good things of it, which few did entertain, and (because their description was somewhat dark) most despised and defamed them for untruths.

In like manner, and much more may this my writing be lightly esteemed of, because I do not set down long and costly processes, but only according to truth, and in simplicity do labor to serve my neighbor, which does not sound well in the ears of the proud world, which rather tickle and load themselves with vain, and unprofitable processes, than harken unto the truth; and it is no wonder, that God suffers such men, which only look after high things, and despise small things, to be held in Error.

Why do we look to get our Medicines by troubling our brains, & by subtle and tedious works, whereas God through simple nature does teach us otherwise? Were it not better to let simple nature instruct us? Surely if we would be in love with small things, we should find great ones. But because all men do strive only for great and high things, therefore the small also are kept from them; and therefore it would be well, that we could fancy this maxim, that also things of small account can do something, as we may see by Tartar and despicable Antimony, and not only so many coals, glasses, materials, and the like, but also the precious time would not be wasted so much in preparing of medicaments: for all is not gold that glitters, but oftentimes under a homely coat some glorious thing is hid; which ought to be taken notice of.

    Some may object why do I teach to join the Antimony first with the Tartar by the help of common water before its fermentation with the wine: whether it would not be as good to put it in of itself in powder, or to dissolve it with spirit of salt (which would be easier to do than with Tartar) and so let it work? To which I answer, that the working wine or drink, receives no metallical calx or solution, unless it be first prepared with tartar or spirit of wine. For although you dissolve Antimony, or any other metal or mineral in spirit of salt, or of vitriol, or of salt nitre, or any other acid spirit, and then think to let it work with wine or any other drink, you will find that it does not succeed; for the acid spirit will hinder the fermentation, and let fall the dissolved metals, and so spoil the work; and besides, Tartar may be used among all drinks, and does more agree with one's taste and stomach, than any corrosive spirit.

    In the same manner as was taught of Antimony, other minerals and metals also may be fitly joined with wine or other drink, and the use of such Antimonial wine is this, viz. that it be drank at meals and betwixt meals like other ordinary drink to quench thirst, but for all that, it must not be drank in a greater quantity, than that Nature be able to bear it. For if you would drink of it immoderately, it would excite vomits, which ought not to be, for it is but only to work in an insensible way, which if it be done, it preserves not only the body from all diseases proceeding from corrupted impure blood, as the Plague, Leprosie, Pox, Scurvy, and the like, but by reason of its hidden heat, whereby it does consume and expel all evil and salt humours (as the Sun dries up a pool,) by sweat and urine, and so does unburden the blood from all such sharp and hurtful humours, & etc. It does not only cure the above said diseases, but also all open sores, ulcers, fistulas, which by reason of the superfluity of salt humours can admit of no healing, and it does dispatch them in a short time in a wonderful manner, and so firmly that there is no relapse to be feared.

This drink is not only good for the sick, but also for the whole (though in a smaller quantity) because that it wonderfully cleanses the whole body, and you need not fear the least hurt either in young or old, sick or healthy. And let no man stumble at it, that many ignorant men do defame Antimony and hold it to be poison, and forbid it to be used, for if they knew it well, they would not do it; but because such men know no more, than what they get by reading, or by hear—say, they pronounce a false sentence; and it might be replied unto them, as APOLIES did to the Shoe maker; NE SUTOR ULTRA CREPIDAN: but what shall we say? NON OMNIS FERE OMNIA TELLUS. When an Ass after his death does rot, out of the carcass grows Beetles, which can fly higher than the Ass from whence they came; in the like manner, we wish it may fare with the haters of royal Antimony, viz. that their posterity may get seeing eyes, and what they know not, they may forbear to despise and scoff at.

I must confess, that if Antimony be not well prepared, and besides, be indiscreetly used by the unskillful, that it may prejudice a man in his health, which even the vegetables also may do. But to reject it because of the abuse, would be a very unwise act: if perchance a child should get into his hand a sharp-edged knife, and hurt himself or others, because it does not understand how to use a knife, should therefore the use of a knife be rejected and forbidden to those that are grown up and know how to use it? Good sharp tools make a good workman; so, good quick working and powerful medicines make a good physician; and the sharper the tool is, the sooner a stone-carver or other craftsman may spoil his work by one cut which he does amiss: which also must be understood of powerful medicines, for if they be used pertinently, in a short time more good may be done with them, than with weak medicaments in a long time. Now as a sharp tool is not to be handled but by a good workman, so likewise a powerful medicine ought to be managed by an understanding and conscientious physician, who according to the conditions of the person, and the disease, knows to increase or abate the strength of the medicine, and not by such a one, as does minister it ignorantly without making any difference at all.

Let no man marvel, that I ascribe such great virtues unto Antimony, it being abundantly enriched with the PRIMUM ENS of gold. If I should say ten times as much more of it, I should not lie. Its praise is not to be expressed by any man's tongue; for purifying of the blood, there is no mineral like unto it; for it cleanses and purifies the whole man in the highest degree, if it be well prepared first, and then discreetly used. It is the best and next friend to gold, which by the same also is freed and purified from all addition and filth, as we said even now, of man. Every Antimony for the most part agrees with gold and its medicine; for out of Antimony, by the cleansing Art may be made firm gold, as in the fourth part shall be taught, and which is more, by a long digestion a good part of, the same is changed into gold. Whereby it is evident, that it has the nature and property of gold, and it is better to be used for a medicine, than gold itself, because the golden virtue is as yet volatile in this, but in the other is grown fixed and compacted, and may be compared to a young child in respect of an old man. Therefore, it is my advice, that in Antimony medicine should be sought, and not to trifle away time and cost in vain and useless things.

Further note, that if you desire to contract nearer together the virtue of Antimony or any other mineral or metal, as above was taught to be done with the Tartar, you must by exhalation of the superfluous moisture in Balneo, reduce the solution to a honey thick liquor, and pour spirit of wine upon it for to extract, and within few days it will be very red; then pour it off and pour on other, and let this likewise extract: continue this proceeding with shifting the spirit of wine, till the spirit of wine can get no more Tincture; then put all the colored spirit of wine together into a glass with a long neck, and digest it so long in a warm Balneum, till the color or best essence of Antimony be separated from the spirit of wine, and settled to the bottom like a blood red thick fat oil, so that the spirit of wine is turned white again; which is to be separated from the fair and pleasant oil of Antimony, which is made without any corrosive, and is to be kept as a great treasure in physick. The spirit of wine retains somewhat of the virtue of Antimony, and may be used with success of itself both inwardly and outwardly. But the Tincture as a Panacea in all diseases acts its part with admiration, and as here mentioned of Antimony, so in the same manner all metals by the help of Tartar and spirit of wine may without distilling be reduced into pleasant and sweet oils, which are none of the meanest in Physick: for every knowing and skillful Chymist will easily grant, that such a metallical oil, as without all corrosives out of the gross metals is reduced into a pleasant essence, cannot be without great and singular virtue.

How to Make Good Spirit and Oils out of Pearls, Corals, Crabs—eyes, and other light solable Stones of Beasts and Fishes.

Take to one part of pearls or corals (made into fine powder) three or four parts of pure Tartar, and so much water as will dissolve the Tartar by boiling; put the corals, Tartar and water together into a glass body, which must stand in sand, and give it so strong a fire, that the water boils in the glass body with the Tartar, and may dissolve the corals. (This solution may be done also in a clean earthen pot that is glazed, and the evaporated water must be supplied with other, as above was taught to be done with the metals.) The corals being dissolved, let them cool, filtrate the solution, and abstract all the moisture from it in Balneo, and there will remain a pleasant honey-thick liquor, which may be used in Physick either of itself, or else once more extracted with spirit of wine and purified, or else distilled, as you please.

The extract or Tincture is better than the liquor, and the spirit is better than the extract or tincture: and all three may well and safely be used; they strengthen the heart and brain; especially those which are made of pearls and corals, they expel the urine and keep the body soluble. Those of crab's eyes and of perches and other fishes open and dense the passages of the urine from all slime and impurity, and they powerfully expel the stone and gravel in the reins and bladder.

N. B. The distilled spirit of corals being well rectified, is good for the Epilepsy, Melancholy, and Apoplexy. It expels and drives out all poison by sweating, because it is of a golden nature and quality, whereof in another place more Shall be said.

To Distil a Spirit out of Salt of Tartar and Crude Tartar.

If you take a like quantity of crude Tartar and of salt of Tartar, and dissolve it with clean water, and evaporate the water still skimming it, till no more skin rises, and then let it cool, there will shoot white crystals, which being distilled as common Tartar, they will yield a purer subtler and pleasanter spirit, than the crude Tartar does, in all to be used as above has been taught of the simple spirit of Tartar: therefore it is needless here to describe its use. Before you distil a spirit thereof, you may use them instead of TARTARUS VITRIOLATUS for purging, they will cause gentle stools, and drive also the urine and stone, and are not unpleasant to take. The dose is from 1/2 to 1 ounce in waters fit for your purpose. This salt dissolved with water purifies metals (if they be boiled therein) and makes them fairer then common Tartar does.

How to get a powerful Spirit out of the Salt of Tartar, by the help of pure Sand or Pebble-stones.

In the first part of this book I taught how to make such a spirit, but because the materials, which are to be distilled in that furnace must be cast upon quick coals, whereby the remainder is lost, and that also not everyone has the convenience to set up a furnace that requires more room than this here does: therefore I will set down how it may be got with ease in this our present furnace, without the loss of the remainder, which is not inferior to the spirit itself. And it is done thus:

Make a fair white salt of calcined Tartar by dissolution, filtration and coagulation, pulverize that salt in a warmed mortar, and add to it a fourth part of small pulverized crystal or flints or only of fine sand, washed clean, mix it well, and cast one spoonful thereof at once into your red-hot vessel (which must be made of earth) and so cover it, and the mixture as soon as it is red hot, will rise and boil (as common Allome does, when it comes to a sudden heat) and yield a thick white heavy spirit; and when it ceases to come forth, then cast in another spoonful, and stay out the time of its settling, and then another part again, till all your mixture be cast in. When no more spirit goes forth, then take off the lid from the distilling vessel, and with an iron ladle take out that which stayed behind, whilst it is yet red-hot and soft, and it will look like unto a transparent clear white fusible glass, which you must keep from the air, for it will dissolve in it, till I teach you what you are to do with it.

    The spirit which came over, may either be kept as it is, or else rectified PER ARENAM in a glass retort, and used in Physick; it is clean of another taste than the spirit of common salt or vitriol, for it is not so sharp; it smells of the flints after a sulphureous manner, and tastes urine-like, and it is very good for those that are troubled with the gout, stone and Tisick: for it provokes urine and sweat mightily, and (because it cleanses and strengthens the stomach) it also makes one have a good appetite to his Victuals. What else it can do is unknown to me as yet, but it is credible that it may act its part in many other diseases, which is left free for everyone to try. In my opinion (since the spirit of the salt of Tartar is good to be used of itself for the stone, and that here it is strengthened by the sand, which have the signature of the stone of the Microcosme) there is hardly any particular medicine, which can go beyond it, but I leave everyone to his own opinion and experience. Externally used, it quenches inflammations and makes a pure skin, & etc. The remainder, which I bid you keep, and looks like a transparent clear glass, is nothing else but the most fixed part of the salt of Tartar and flints, which joined themselves thus in the heat, and turned to a soluble glass, wherein lies hid a great heat and fire. As long as it is kept dry from the air, it cannot be perceived in it: but if you pour water upon it, then its secret heat will discover itself.

If you make it to fine powder in a warm mortar, and lay it in a moist air, it will dissolve and melt into a thick fat oil, and leave some faeces behind. This fat liquor or oil of flints, sand or crystal may not only be used inwardly of itself, but also serves to prepare minerals and metals into good medicines, or to change them into better by Chymical art. For many great secrets are hid in the contemptible pebble or sand; which an ignorant and un-expert man (if they were disclosed to him) would hardly believe: for this present world is by the devil's craft so far possessed with cursed filthy avarice, that they seek for nothing but money, but honest and ingenious sciences are not regarded at all; and therefore, God does close our eyes that we cannot see what lies before them, and we trample upon with our feet. That worthy man PARACELSUS has given us sufficient information to understand, when he says in his book (containing the vexations of Alchymists) that many times a despicable flint cast at a Cow is worth more than the Cow; not only because that gold may be melted out of it, but also that other inferior metals may be purified thereby, so that they are like unto the best gold and silver in all trials; and although I never got any great profit by the doing of it, yet it does suffice me that I have seen several times the possibility and truth thereof, which in its proper place likewise shall be taught.

    This liquor of flints is of that nature toward the metals, that it makes them exceeding fair, but not so, as women do scour their vessels of tin, copper, iron, & etc. with lye and small sand, till all filth be scoured off, and that they get a bright and fair gloss: but the metals must be dissolved therein by Chymical art, and then either after the wet or dry way digested in it for its due time; which PARACELSUS calls to go into the mother's womb, and be born again: if this be done rightly, then the mother will bring forth a pure child. All metals are engendered in sand or stone, and therefore they may well be called the mother of metals, and the purer the mother is, the purer and sounder child she will bear, and among all stones there is none found purer than the pebble, crystal or sand, which are of one nature (if they be simple and not impregnated with metals). And therefore, the pebble or sand is found to be the fittest bath to wash the metal withal. But he that would take this bath to be the Philosophers Secret Menstruum, whereby they exalt the King unto the highest purity, would be mistaken; for their Balneum is more friendly to gold because of its affinity with it than with other metals, but this does easier dissolve other metals than gold. Whereby it is evident, that it cannot be BERNARD his fountain (BERNHARDI FONTINA) but must be held only to be a particular cleanser of metals. But omitting this, and leaving it to the further practice and trial of those that want no time nor convenience for to search what may be done with it, let us take notice of the use of this liquor in physick, for which uses sake this book is written. That which has been said, was only done to that end, that we may observe, that we must not always look upon dear and costly things, but that many times even in mean and contemptible things (as sand & pebbles) much good is to be found.

How to Extract a blood-red Tincture with Spirit of Wine out of the Liquor of Pebble-stones.

If you extract a tincture out of pebble-stones, for use in Physick or in Alchemy, then instead of the white take a fair yellow, green or blue pebble or flint, whether it hold fixed or volatile gold, and first with salt of tartar distil the spirit thereof; or if you do not care for the spirit, then melt the mixture in a covered crucible into a transparent, soluble and fusible glass, and in a warm mortar make it into fine powder; put this powder in a long necked glass, and pour upon it rectified spirit of wine (it need not be dephlegmed, it matters not if it be pure), let it remain upon it in a gentle warmth, till it turns red (the glass with the prepared pebble or flints must be often stirred about, that the pebble be divided, and the spirit of wine may be able to work upon it) then pour off the colored spirit of wine, and pour on other, and let this likewise turn red: this pouring off and on must be iterated so often, till the spirit of wine gets no more color out of it. All the Tinctured spirit of wine put together, & abstract in a Balneum through a Limbeck from the Tincture which will remain in the bottom of the glass body like a red juice, which you must take out and keep for its use.

The use of the Tincture of Pebbles or Flints in Physick.

This Tincture if it be made of gold, pebbles or sand, is to be held for none of the least medicines, for it does powerfully resist all soluble Tartareous coagulations, in the hands, knees, feet, reins, and bladder; and although in want of those that hold gold, it be extracted but only out of common white pebble, it does act its part however, though not altogether so well as the first. Let no man marvel, that sand or pebbles made potable, have so great virtue; for not all things are known to all; and this Tincture is more powerful yet, if first gold have been dissolved with the liquor of pebbles before the extraction. And let no man imagine that this Tincture comes from the salt of Tartar (which is taken to the preparing of the oil of sand) because that of itself also does color the spirit of wine, for there is a great difference betwixt this Tincture and that, which is extracted out of the salt of Tartar: for if you distil that of the salt of tartar in a little glass body or retort, there will come first a clear spirit of wine, then an unsavory phlegm, and a salt will remain behind, in all like unto common salt of tartar, wherein after its calcining not the least color appears, and because none came over either, it might be questioned where it remained then?

To which I answer, that it was not a true tincture, but only that sulphur in the spirit of wine was exalted or graduated by the corporeal salt of tartar, and so got a red color, which it loses as soon as the salt of tartar is taken from it, and reassumed its former white color: even as it happens also, when the salt of urine, or of harts-horn or soot, or any other like urineous salt is digested with spirit of wine, that the spirit turns red of it, but not lastingly, but just so as it falls out with the salt of tartar, for if by rectification it be separated again from the spirit of wine, each (viz. both the salt and also the spirit of wine) does recover again its former color, whereby it appears, that (as above said) it was not a true tincture. He that will not believe it, let him dissolve but 1 ounce of common white salt of tartar in 1 lb. of spirit of wine, and the spirit will turn as red of it, as if it had stood a long time upon several pounds of blue or green calcined salt of tartar; and if I had not tried it myself several times, I should have also been of that opinion: but because I found it to be otherwise, therefore I would not omit to set down my opinion: though I know I shall deserve small thanks of some, especially of those which rather will err with the greater number, than to know and confess the truth with the less number. However, I do not say, that the supposed tincture of the salt of Tartar is of no virtue or useless; for I know well enough that it is found very effectual in many diseases: for the purest part of the salt of Tartar has been dissolved by the spirit of wine, it being thus colored thereby, and therefore that tinctured spirit of wine may very fitly be used. But as for the Tincture, which is extracted out of the prepared pebbles, it is clean of another condition: for if you abstract the spirit of wine from it, though it also comes over white, yet there remains a deep tinctured salt, whose color is lasting in the strongest fire, and therefore may be counted a true Tincture.

How by the help of this Liquor out of Gold its Red color may be Extracted so that it remains White.

This, oil or liquor of pebbles is of such a condition, that it does precipitate all metals which are dissolved by corrosives, but not after that manner as the salt of Tartar does; for the calx of metals which is precipitated by this liquor;

(because the pebbles mingle therewith) is grown much heavier thereby, than if it had been only precipitated with salt of Tartar.

For example, dissolve in AQUA REGIA as much Gold as you please, and pour of this liquor upon it, till all the Gold fall to the bottom like a yellow powder, and the solution turns white and clear, which you must pour off, and edulcorate the precipitated Gold with sweet water, and then dry it (as you were taught to do with the AURUM FULMINANS) and you need not fear that it will kindle and fulminate in the drying, as it used to do, when it is precipitated with salt of Tartar or spirit of urine, but you may boldly dry it by the fire, and it will look like yellow earth, and will weigh as heavy again as the Gold weighed before the solution; the cause of which weight is the pebble stones, which precipitated themselves together with the Gold. For the AQUA REGIA by its acidity has mortified the salt of Tartar, and robbed it of its virtues so, that it could not choose but let fall the assumed pebbles or sand; on the other side, the salt of Tartar which was in the liquor of pebbles, has annihilated the sharpness of the AQUA REGIA, so that it could not keep the dissolved gold any longer, whereby both the gold and the pebbles are freed from their dissolver.

This edulcorated and dried yellow powder put into a clean crucible, and set it between live coals, that it begins to be red hot, but not long, and the yellow will be changed into the fairest purple color, which is pleasant to behold, but if you let it stand longer, then the purple color vanishes, and it turns to a brown and brick color: and therefore if you desire to have a fair purple colored gold, you must take it off from the fire, as soon as it is come to that color, and let it not stand any longer, else it loses that color again.

This fair gold—powder may be used by the rich (which are able to pay for it) from 1/2 to 1/8 ounce, in convenient vehicles; and in all diseases, where sweating is needful: for besides the provoking of sweat, it comforts not only the heart, but also by the virtue of the pebble it expels the stone in the reins and bladder (if it be not grown to the height of hardness) like sand together with the urine: so that it may be safely used as well to prevent, as to cure the plague, gout and stone.

How to make further out of this purple colored gold a soluble Ruby for medicinal use, shall be taught in the fourth part: for in regard that it must be done by a strong fire in a crucible, it does not belong hither, but to its proper place, where other like Medicaments are taught to be made.

If you will extract the color out of this precipitated gold, then pour upon it (before it be, put into the fire to calcine) of the strongest spirit of salt, and in a gentle heat the spirit will dissolve part of the gold, which will be much fairer and deeper in color, than if it had been done with AQUA REGIA: upon this solution pour five or six times as much of dephlegmed spirit of wine, and digest both together its due time, then by the digestion of a long time, part of the Gold will fall out of the solution to the bottom like a fair white powder, which may be reduced with Borax or salt nitre and Tartar; it is white like silver, and as heavy as other gold, and may easily get its color again by the help of Antimony. The residue out of which the white gold is fallen, viz. the spirit of salt mingled with the spirit of wine, must be abstracted from the Tincture, and there will remain a pleasant sour liquor colored by the gold, upon the bottom of the glass body, which is almost of the same virtue, which above has been ascribed to other tinctures of gold. Especially this liquor of gold strengthens the heart, brain, and stomach.

N. B. Sometimes there comes over with the spirit of wine a little red oil, which the strong spirit of salt has separated from the spirit of wine, and it is impregnated with the Tincture of Gold. It is an excellent cordial, few are found like unto it, whereby weak people decayed by sickness or age, may be kept alive a long time, they taking daily some drops of it, who else for want of the HUMIDUM RADICALE, would be forced to exchange their life for death.

Here some body may ask, whether this Tincture is to be counted or taken for a true Tincture of Gold; or whether there be another better to be found?

To which I answer, that although many may hold it to be such, and I myself do call it so here, yet that after due examination it will not prove to be such: for although some virtue is taken from the gold by this way, yet it does still keep its life, though it be grown weak and pale, because it can so easily recover its former sound color by a contemptible mineral: if its true Tincture or soul were gone from it, surely an inferior mineral could not restore it to life, but of necessity there would be required such a thing to do it, which has not only so much, as it has need of for itself, but has a transcendent power to give life unto dead things. As we may see by a man or any sensible beast, that if they have lost their vigor by adversities, in that no life more is perceived in them, yet by medicines fit for the purpose, they may be refreshed, and brought to their former health, so that their former disease appears no more in them; but if their soul be once gone, the dead body can by no medicines be restored unto life again, but must remain dead so long, till he in whose power it is to give and to take life, have mercy upon it. So likewise, it is to be understood of gold, when its color is taken from it, and yet its life is left, which by the help of Antimony, being its medicine, as also by the help of iron or copper can be restored unto it, so that it recovers its former fair colors, so that you cannot see at all, that it ailed anything before. But if its life be gone from the body, it is impossible for any ordinary metal or mineral to restore it to life, but it must be done by such a thing, which is more than Gold itself has been: for even as a living man cannot give life unto a dead man, so but GOD must do it who has created man; so Gold cannot restore to dead Gold, the life which has been taken from it, and how could it then be done by an unfixed mineral? But there is required a true Philosopher for to do it, such a one as has good knowledge of gold and its composition.

Now as we heard that like cannot help its like, but he that shall help, must be more, than he that looks for help from him. Hence it is evident, that the Tincture, whose remaining body (from which it is taken) is still gold, can be no true tincture; for if it shall be a true tincture, it must consist in its three principles, and how can it consist therein, the body from whence it came being yet alive, and possessing invisibly all its three principles? How can a man's soul be taken from him, and yet the body live still? Some will say, that for all that, this may be counted a true tincture, although the body still remains gold, and have kept its life: even as man may spare some blood out of his body, which though it will make him somewhat pale, yet he still lives, and the lost blood may be supplied again by good meat and drink. But what lame and senseless objections are these? Who would be so simple as to think, that a handful of blood may be compared to a man's life? I believe no wise man will do it. Although life goes forth with the blood, yet the blood is not the life itself; else the dead could be raised thereby, if a cup full of it were poured into a dead body; but where was such a thing ever heard or seen? With such groundless opinions some did presume to censure the truth, set down in my treatise DE AURO POTABILI VERO, saying, GEBER and LULLIUS were also of opinion, that a true tincture can be extracted out of gold, the same nevertheless remaining good gold: but it may be asked, what it has lost then for to yield a true Tincture, since it remained good gold? Here nobody will be at home to answer, I doubt. What are the Writings of GEBER or LULLY to me? What they have written I do not despise, they were highly enlightened and experienced Philosophers, and would defend their writings sufficiently, if they were alive: and what I write, I am also able to maintain.

Do those men think, that the writings of GEBER and LULLY are to be understood according unto the bare letter? Show me a tincture of gold which was made by the writings of GEBER or LULLY. If it were so, then every idiot or novice, that could but read Latin, would not only by their writings be able to make the Tincture of Gold, but also the Philosopher's Stone itself, whereof they have written at large; which does not follow, because it is seen by daily experience that the most worldly learned men spent many years, and have been at vast charges, and taken great pains, and studied in their books day and night, and found not the least thing in them.

Now if such Philosophers were to be understood literally, doubtless there would not be so many poor decayed Alchymists. Therefore, the writings of such worthies are not to be understood according to the letter, but according to the mystical sense hid under the letter.

But because the truth is eclipsed in their books by so many seducing and sophistical processes, there will hardly any man be able to pick it out from so many seducements, unless a light from God be given to him first, whereby he may be able to peruse the dark writings of those men, that he knows how to separate the parabolical speeches, from those that are true in the letter itself: or if an honest Godly Chymist by the grace of God in his labors hits upon the right steps, and yet do doubt, whether he be in the right way or not, then by reading of good and true Philosophers books, he may at last learn out of them the firm and constant truth: else hardly any one's desire may be obtained out of their books, but rather after the precious time spent, means and health wasted, a man shall be forced to fall a begging at last.

In like manner, if the true tincture be taken from Copper, the rest is no more a metal, nor by any Art or force of fire can be reduced to a metallical substance.

N. B. But if you leave some tincture in it, then it may be reduced into a brittle gray body, like unto iron, but brittle.

Another way to Extract a good Tincture out of Gold by the help of the Liquor of Sand or Pebbles.

Take of that gold calx (which was precipitated with the oil of sand) one part, and three, or four parts of the liquor of crystals or of sand, mix the gold calx in a good crucible with the liquor, and set this mixture into a gentle heat, so that the moistness may evaporate from the oil of sand which is not easily done; for the pebble or sand, by reason of their dryness keep and hold the moistness, and will not let it go easily; it rises in the pot or crucible, as borax or Allome does when you calcine them; therefore the crucible, must not be filled above half, that the liquor together with the gold may have room enough, and do not run over the pot: and when it rises no more, then strengthen the fire, till the pot is red hot. The mixture standing fast, put a lid upon it, which may close well, that no coals, ashes, or other impurity may fall into it, and give it so strong a fire in a wind furnace, that the liquor together with the gold calx may melt like water; keep it melting so long, till the liquor and gold together be like unto a transparent fair ruby, which will be done in an hours' time or thereabouts; then pour it forth into a clean copper mortar, let it cool, and then make it into powder, and pour spirit of wine upon it to extract, which will look like thin blood: and will prove more effectual in use, than the above described Tincture.

The residue from which the Tincture is extracted, must be boiled with lead, and precipitated and driven off as you do ores, and you will get the remaining gold, which went not into the spirit of wine; but it is very pale and turned like unto silver in color, which if it be melted by

Antimony, it recovers its former color without any considerable loss in the weight. How the melting in crucibles, and boiling of the remaining gold is to be done, shall be more punctually set down in the fourth part. I know several other fine processes, to extract the color easily out of gold; but because the gold must be first made fit for it by melting in a crucible, and that it is not pertinent to speak of that operation here in this second part, therefore it shall be reserved for the fourth, where you shall be informed at large, not only how to prepare Gold, Antimony and other minerals, and make them fit for extraction, but also how to reduce them into a transparent, soluble and fire-proof Ruby (which are none of the meanest medicines) and as it was done here with the gold, so you may proceed likewise with other metals and minerals to extract their colors. And therefore, being needless to describe each metals tincture by itself, all the processes of them shall be disclosed in one, viz. in that of gold. The book would grow too big, if I should describe them severally, which I count needless to do. Let this suffice for this Second part, that we have taught, how to extract out of the gold its color after a common way. Which indeed are good medicines, but for ought I know of no use in Alchemy. But he that seeks a true Tincture out of gold, let him endeavor first to destroy the gold by the universal Mercury, and to turn the inside outward, and the outside inward, and proceed further according unto art, then the soul of gold will easily join itself with the spirit of wine, and come to be a good medicine, whereof more in my treatise DE AURO POTABILI is handled. If one know the CHALYBS of SENDIVOGIUS, which is well to be had, he might with little labor quickly get a good medicine: but because we show ourselves still ungrateful children unto God, therefore it is no marvel, that he withdraws his hand from us, and leaves us in errors.

What further may be done with the Liquor of Pebbles.

Many more profitable things, as well in Alchemy, as in medicine, may be compassed by the oil of sand; as for example, to make fair painting colors out of metals, which abide in all elements. Also, to frame all sorts of transparent hard stones out of crystals, which in beauty are like unto the natural, yea fairer sometimes; also, how to make many fair Amauses or Enamels and the like profitable arts: but they belong not to this second part, shall be reserved for the fourth, where all such shall be taught very punctually with all the circumstances thereunto relating.

How by the help of this Liquor to make Trees to grow out of Metals, with their Colors.

Although this process in Physick may be of no great use: yet in regard that to a Chymical Physician it gives good information of the condition of natural things, and their change. I thought it not amiss to set it down here.

Take of the above described oil made of sand, pebbles or crystals as much as you please, mix therewith a like quantity of the lixivium of Tartar, shake both well together, so that the thick liquor may not be perceived in the lixivium, but be thoroughly incorporated therewith; both being turned to a thin solution, and then your water is prepared, wherein the metals grow.

The metals must first be dissolved in their proper corrosive MENSTRUUM, and the MENSTRUUM must be quite abstracted from thence again, but not too near, that the calx of the metal may not grow red-hot, whereby its growing virtue would be taken from it. Then take it out of the little glass-body, and break in pieces about the bigness of a pulse, and put them in the above described liquor in a clear bright glass, that the growing of the metals may be discerned through it; and as soon as the prepared metals are taken out of the glass body, they must be kept from the air, else they lose their growing virtue. Therefore, thus dry they must be broken in pieces, and laid in the bottom of the glass (wherein the liquor is) a fingers breadth one from another asunder, and must not be laid together on a heap. The glass must stand still in a quiet place, and the metal will presently swell in it, and thrust forth some bulks, out of which branches and twigs do grow, so finely, that one shall admire at it; and let none think that this growing serves only to please the eye, for some special thing is hidden in it; for all sand or pebbles, although they be white, invisibly contain a hidden tincture or golden sulphur, which none without experience will be able to believe; If for a time you digest the pure filings of lead in it, gold will come to stick to the outside thereof (which gold may be washed off with water) and the lead will look as if it were gilded. Which gold came from nowhere else but from the sand or pebbles, although they were white and clear, so that it could not be perceived in them. It shows also its meliorating virtue, when the metals do grow therein, and for a certain space of time are digested therewith. For it may be seen apparently, that the metals in the growing do increase from this liquor, and attract what is for their turn; which hence also may be perceived, that when but as much as the bigness of a pea grows therein, it will grow twice or thrice as big, which is worthy to be considered. Also, the pebbles or sand—stones are the natural matrixes of metals, and there appears a great Sympathy between them, especially between the unripe metals and them; as if nature should say to such raw or unripe metals, return into your mother's womb, and stay there the due time, till you have attained there to perfect ripeness, for you were taken thence too soon against my will. Further, out of this liquor there may be made a good borras to reduce the metals thereby. There may be made also with this liquor fair glased and firm colors upon earthen vessels like unto Porcelain or China. Also by boiling it with water, a tender impalpable snow—white earth may be precipated out of it, whereof there may be made vessels like unto Porcelain.

Many other useful things may be brought to pass thereby in mechanical businesses, needless here to relate.

Also, unripe and volatile minerals may be fixed and ripened thereby, so that not only they may be the fitter to be used in Physick, but also the volatile gold and silver contained in them may be saved thereby, whereof more in the fourth part.

N. B. Hither belongs also the process of the spirit of lead, Virgins-milk and Dragons blood.

Of the Spirit of Urine and of the volatile Spirit of Salt Armoniack.

Out of Urine or Salt Armoniack, a powerful and penetrating spirit may be made several ways, which not only is to be used in physick for many diseases, but is also found very useful in mechanical and chymical operations, as follows.

Take of the urine of sound men living chast, gather a good quantity together in a wooden vessel, let it stand for its time to putrefy, and distil a spirit thereof, which afterward in a great glass retort with a wide neck must be rectified from calcined tartar, and still that which comes over first, may be saved by itself, and so the second and third also, the strongest may be used for the preparing of metallical medicines, and the weaker for a medicine alone by itself, or else mingled with fit vehicles. The salt which in the rectification comes over with the strongest spirit; may be put to the weakest, to make it the stronger, or else it may be saved by itself in a good strong glass.

But because the spirit of urine is tedious to make, therefore I will show, how to get it easier out of salt Armoniack. The preparation is thus.

Take of salt Armoniack, and LAPIS CALAMINARIS, and make each by itself into powder, and then mix them together, and cast of it into the red-hot vessel at once no more than 1 or 1/2 ounce. Unto the vessel there must be applied a great receiver: for the spirit goes with such a force and power, that it was impossible to distil it in a retort without danger or loss, for I broke more than one receiver with it, before I did invent this instrument. The spirits being well settled in the receiver, cast in more of your mixture; this continue so long till all your matter is cast in; then take off the receiver, and pour the spirit into a strong glass, which must not be well closed at the top, but not with wax and a bladder, because it softens the wax, and does penetrate through the bladder; but first stop it with paper, then melt Lacca or sulphur, and pour it upon it, so that it comes to be very well closed, and then it will not be able to exhale, or you may get such glasses made, as in the fifth part shall be taught, to keep all the subtle spirits in them, for more security. And this spirit, if no water has been mixed with it in the receiver, needs no rectifying: but he that will have it stronger yet; may rectify it by a glass retort, and so keep it for use.

And this is the best way to make a strong spirit out of salt Armoniack: the same may be done also, by taking of filed Zinck, instead of LAPIS CALAMINARIS: also by adding of salt of tartar, salt made of the Lee of wood ashes, unquenched lime, and the like: but the spirit is nothing near so strong (although all those things may be done with it, that are done with the former) as that which is made with LAPIS CALAMINARIS or Zinck.

The process or the manner of making it, is thus:

Take 1 lb. of salt armoniack made into powder, and as much of salt of tartar, mix both together by the help of a lye made of tartar, or only with common water, so that all come to be like a pap, and cast in one spoonful thereof at once, into the distilling vessel, then cast in more till you have spirit enough.

N. B. The salt of tartar may also be mixed dry with the salt armoniack without any lye or water, and so distilled: but it is not so good, as when the mixture is tempered with lye or water: for if it be cast in dry, the spirit will come over in the form of a volatile salt: but if the mixture has been moistened, then most part thereof will come over like a fiery burning spirit. In like manner, also the mixture of Lime and salt armoniack may be tempered moist, and it will yield more spirit than if it be distilled dry.

It may be asked: why LAPIS CALAMINARIS, Zinck and unquenched lime, calcined tartar, salt of potashes, fixed salt nitre or the like things prepared by the fire, must be added unto salt armoniack, and whether it be not good to add some bolus, or other earth (as usually is done to other salts) and so distil a spirit of it? To which I answer, that there are two sorts of salt Armoniack, viz. a common acid salt, and a volatile salt of urine, which without mortifying of one of them, cannot be separated: for as soon as they feel the heat, the volatile salt of urine carries the acid salt upwards, and they both together yield a sublimate, of the same nature and essence with common salt armoniack which is not sublimed, only it is purer than the common. And no spirit would come over from it, if it should be mingled with bole, brick, dust, sand, or any other strengthless earth, and so distilled, but the whole salt as it is of itself (leaving its earthly substance behind) would sublime thus dry: but that it falls out otherwise with the LAPIS CALAMINARIS (which is also like an earth) so that a separation of the salts is wrought thereby, and a volatile spirit comes over; the reason is, that the LAPIS CALAMINARIS and Zinck are of such a nature, that they have a great affinity with all acid things, and do love them, and are loved by them likewise (whereof some mention has been made in the first part) so that the acid salt sticks to it in the warmth, and unites itself with it, and the volatile salt is set free, and distilled into a subtle spirit; which could not have been done, if the acid salt had not been kept back, by the LAPIS CALAMINARIS or Zinck. But that a spirit is distilled off by addition of fixed salts; the reason is that fixed salts are contrary unto acid salts, and (if they get the upper hand) do kill the same, and rob them of their strength, whereby those things which are mixed with them are freed from their bond: and so it falls out here with salt armoniack, that by addition of a vegetable fixed salt, the acidity of the salt armoniack is killed; the salt of urine, which formerly was bound therewith, gets its former freedom and strength, and being sublimed turns into a spirit. Which could not have been done, if common salt had been added to the salt armoniack instead of salt of tartar; for the salt of urine would thereby (as by a far greater enemy be killed and kept back, so that it could yield no spirit). I thought fit to give notice hereof to the ignorant (not for those, who knew it before) and to the unknowing it will do much good, and that they may have a light for other labors: for I have many times seen, and see it still by daily experience, that the most part of vulgar Chymists, whatsoever they do (having got it either by reading, seeing, or hearing) they hurl it over like botchers, and are not able to give any solid reason, why this or that must fall out in such or another manner, not laboring to find out the natures and conditions of salts, minerals, and other materials, but contenting themselves only with the Receipt[14], saying this or that Author has written so, and therefore it must be so, whereas many times such books are patched up out of all sorts of authors. And those that stick to so many books, will hardly ever come to get any good, but are led out of one Labyrinth into another, spending their life miserably in watching and cares: but if they would first seriously consider things, and learn to know

---

[14] Most likely "recipe" or "medical prescription." -PNW

nature, and then take their work in hand, then they would sooner attain unto true knowledge; and so much of this matter by the way. I hope that he that has been in error will be pleased with it, and the knowing will not grudge us having imparted it to the ignorant.

That which remains after the distillation is done, is also good for use; if the addition has been of salt of tartar, a melting powder may be made of it, to reduce metals. Of LAPIS CALAMINARIS or Zinck, yields PER DELIQUIUM a clear, white, and heavy sharp oil, for the sharper part of salt armoniack, which did not turn to spirit, has dissolved the LAPIS CALAMINARIS, and is almost of the same virtues for external use in Chirurgery with that, which above in the first part which was taught to be made out of LAPIS CALAMINARIS, and spirit of salt, save only that this in the distilling does not yield so strong a spirit as the other, but only yields a sharp sublimate.

Of the Virtue and Use of the Spirit of Salt Armoniack.

This spirit is of a sharp penetrating essence, and of an airy, moist, and warm nature; and therefore, may with credit be used in many diseases, 8, 10, 12 (more or less) drops thereof used in a convenient vehicle, do immediately penetrate all the body over, causing sudden sweating, opening the obstructions of the spleen, and dispersing and expelling many malignities by sweat and urine, it cures the quartane, collick, the suffocation of the Matrix, and many more diseases.

In brief, this spirit is a safe, sure, and ready medicine to disperse and expel all tough, gross and venomous humours. Also, this spirit acts his part externally, quenching all inflammations, curing the Erysipelas and Grangrene; it allays the pains of the gout, clothes being dipped in it and applied: and although it draws blisters, it matters not; laid to the pulse, it is good in ardent fevers, it assuages swellings and pains; disfuses congealed blood, helps strained limbs, and benumbed nerves: only smelled unto, it cures the megrim, and other Chronical diseases of the head: for it dissolves the peccant matter, and evacuates it through the nostrils; it restores the lost hearing, being externally laid on with a little instrument fit for the purpose. Also in the obstructions of women's courses applied by a fit instrument in a spiritual way, opens presently, and cleanses the womb, and makes women fruitful, etc. Mingled with common water, and held in the mouth, assuages the toothache, proceeding from sharp humours which are fallen in the teeth. A little of it applied in a glister, kills the worms in the body, and allays the colick.

This spirit may also further be used to many other things, especially by means thereof many precious and effectual medicaments may be made from metals and minerals, whereof some shall be described as follows.

N. B. There is yet another matter, which is found everywhere and always, and is to be got by everyone without distillation and charges, and is as good for the above said diseases, as the distilled spirit, and if all men knew it, there would not be found everywhere so many sick people, nor so many Doctors and Apothecaries.

To Distil a Blood Red Oil of Vitriol by the help of the Spirit of Urine.

Dissolve Hungarian or other good vitriol in common water, and let it run through a filtering paper, pour of this spirit upon it so much, till all the green is vanished, and the water be made clear, and a yellow sulphur be settled: then pour off the clear, and the rest which is muddy, pour together in a FILTRUM, that the moisture may run off, and the earth of the vitriol remain in the paper, which you must dry, and distil to a blood-red oil, which will open the obstructions of the whole body, and perfectly cure the epilepsie. The clear water must be evaporated dry, and there will remain a salt, which being distilled, yields a wonderful spirit. Before it be distilled, it is a SPECIFICUM PURGANS, whereof 8, 10, 12, to 24 grains taken, may safely be used in all diseases.

The Tincture of Vegetables.

Spices, seeds or flowers being extracted therewith and digested and distilled, the essence of them will come over with it, in the form of a red oil.

Vitriol of Copper.

If you pour it upon calx of copper, made by often heating the copper red hot and quenching it again, it will within an hours' time extract a fair blue color, and having dissolved as much thereof as it can, pour it off and let it shoot in a cold place, and you will get a fair sky colored vitriol, a small quantity whereof will cause strong vomits;

the rest of the vitriol remains a blue oil, good to be used in ulcers.

The Tincture of Crude Tartar.

If you take common crude tartar, and pour of this spirit upon it, and set it in digestion, the spirit will extract a blood-red tincture, and if the spirit be abstracted from it, there will remain a pleasant red oil, of no small virtue and power.

To Make the Oils or Liquors of Salts.

This spirit also dissolves crystals and other stones, they being first dissolved, and precipitated and reduced to impalpable powders, turning them into oils and liquors, good to be used in Alchemy and Physick.

To Precipitate all Metals with it.

Any metal being dissolved in an acid spirit may be precipitated better and purer therewith, than with the liquor of the salt of tartar; for AURUM FULMINANS which is precipitated with it fulminates far stronger than if it were done with oil of tartar.

R. Some juice of Lemon and mix it with the solution of gold, before it be precipitated, and then not all the gold will precipitate, but some of it will remain in the solution, and in time many small green stones (not unlike unto common vitriol) will appear; which in a small dose will purge all noxious humours.

The Oil and Vitriol of Silver.

If you dissolve silver in AQUA FORTIS, and pour so much of this spirit into it till it ceases to make a noise, some of the silver will precipitate in the form of a black powder, the rest of the silver remains in the liquor: abstract the phlegm from it in Balneo, till it gets a skin at the top, and then set into a cool place, there will grow white crystals in it, which being taken out and dried are a good purge in madness, dropsie, fevers and other diseases, safely and without danger to be used to young and old. The rest of the liquor which did not crystallize may be extracted with spirit of wine, and the faeces being cast away the extraction will be pleasanter. The spirit of wine abstracted from it, there will remain a medicine of no small value in all diseases of the brain.

To Extract a Red Tincture out of Antimony or Common Sulphur.

Boil sulphur or Antimony made into powder in a Lixivium of salt of tartar, till it turns red, and pour this spirit upon it, and distil gently in a BALNEUM, and there will come over a fair tincture with the volatile spirit, silver anointed therewith will be guilt, though not lastingly. It serves for all diseases of the lungs.

How to ripen Antimony and common Sulphur, so that several sorts of such Smells, as Vegetables have, arise from Thence.

Dissolve Antimony or Sulphur in the liquor of pebbles or sand, coagulate the solution to a red mass; upon this mass pour spirit of urine, and let it extract in a gentle warmth. The spirit being colored red, pour it off, and pour on other spirit, let it extract likewise, and this you must iterate so often, till the spirit will extract no more tincture; then pour all the extracts together and abstract the spirit of urine from it in Balneum through a limbeck, and there will remain a blood red liquor, and if you pour upon this spirit of wine it will extract a fairer tincture then the former was, leaving the faeces behind, and this tincture smells like garlic: and if it be digested three or four weeks in a gentle warmth, it will get a very pleasant smell, like unto the yellow prunes or plums: and if it remain longer yet in digestion, it will get a smell not inferior to musk and amber; This tincture having been digested a long time, and got several smells, is not only notably by the fire increased in pleasantness of smell and taste, but also in virtue: for so many and various sweet smells are perceived in it, that it is to be admired, which variety and exaltation proceeds only from the pure and ripening spirit of urine, for there is hid in it a fire, which does not destroy but preserves and graduates all colors, whereof in another place more shall be said.

   N. B. Betwixt the spirit of urine and the animal and mineral Copper their appears a great sympathy; for it does not only love copper above all other metals, and mingles easily with it, and makes it extraordinarily fair, and of good use in Physick, but it prepares it also to such a medicine, whereby all venerous sores (both inward and outward use) how deep so ever they took root in the blood, without the use of any other medicaments, are perfectly cured; it makes fruitful and barren, according as it is used; it cleanses the matrix, hinders the rising thereof, and miraculously furthers woman's courses that have been stayed, above all other medicaments of what name so ever.

If this spirit be mingled with the volatile (but not corrosive) spirit of vitriol or common salt there will come a salt out of it, which is inferior to none in subtleness, and useful both in Alchemy and Physick.

N. B. The liquor of the salt of tartar, and the spirit of wine do not mix without water, this being the mean partaking of both of their natures, and if you add unto it spirit of urine it will not mingle but keep its own place: so that these three sorts of liquors, being put in the same glass, and though they be shaked ever so much will not incorporate for all that: the liquor of the salt of tartar keeps to the bottom, next to it will be the spirit of urine, and on the top of that is the spirit of wine: and if you pour a distilled oil upon it, that will keep uppermost of all, so that you may keep four sorts of liquors in one glass, whereof none is mingled with the other.

Although this be of no great profit, yet it serves to learn thereby the difference of spirits.

Of the Spirit and Oil of Harts-horn.

Take Harts-horn, cut it with a saw into pieces, of the bigness of a finger, and cast in one at a time into the aforesaid distilling vessel, and when the spirits are settled, then another, and continue this until you have spirits enough: and the vessel being filled with the pieces that were cast in, take them out with the tongs, and cast in others, and do this as often as is needful. The distilling being finished, take off the receiver, and pour into it dephlegmed spirit of wine, which will cleanse the volatile salt; pour the oil with the spirit and volatile salt through a filtering paper made wet first and lying in a glass funnel, and the spirit of wine together with the spirit of Harts—horn and the volatile salt will run through the paper, and the blackish oil will stay behind, but it must quickly be poured out, else it will pass through after them. The spirit together with the volatile salt rectify through a retort, and the best part of the spirit will come over together with the spirit of wine and volatile salt; and when the phlegme is coming, take of the spirit, which is come over, that the naughty phlegm may not come amongst it; keep it well, for it is very volatile, the oil may be mingled with salt of tartar, and rectified by a glass retort, and so it will be clear; if you will have it fairer, you must rectify it with spirit of salt.

The first, which is done with salt of Tartar, is of more virtue; it cures the Quartane, and provokes sweat extremely, cures all internal wounds and pains, which were caused by falls, blows, or other ways: 6, or 8, 10 to 20 drops of it taken in wine and sweated upon it in the bed. The spirit is very good for all obstructions of the whole body, from 1/24 to 1/8 ounce therefore taken in a fit vehicle, provokes urine, and forces down women's courses, it cleanses the blood and makes sweat mightily. In the Plague, Pox, Leprosie, Scurvy, MELANCHOLIA HYPOCHONDRIACA, malignant Fevers, and the like where sweating is necessary, it proves a rare medicine.

To Make the Spirit of Man's Hair an Excellent Medicine.

After the same manner, you may make spirits out of all kinds of horns and claws of beasts: but since because of their ill smell the use of them is not liked (although in several heavy diseases, as in the fits of the mother and Epilepsie, they do admirably well) therefore I will acquiesce. However, it is worth observing that the spirit made of man's hair is not to be rejected in metallical operations, for it dissolves common sulphur, and reduces it into a milk, which by further ripening may be turned into blood, the like whereunto no spirit can do. The same spirit may also of itself, without addition of sulphur be fixed into a ruby; but that which is ripened with sulphur is the better; and if it be brought so far by the fire, that it has lost its stink, and be made fixed than it will be sufficient to pay for the pains and coals bestowed upon it.

N. B. Hither belongs the Process to pour dissolved metals upon filed harts—horn, and so to distil them.

Of the Oil of Amber.

Amber yields a very pleasant oil and of great virtue especially the white Amber: the yellow is not so good, and the black is inferior to this; for by reason of its impurity it cannot be well used inwardly; and there comes over also along with it a volatile salt and an acid water, which must be separated; the water (for ought that I know) is of little virtue; the salt if it be sublimed from the salt of Tartar and purified, is a good diuretic, and in the Stone and the Gout, may successfully be used both inwardly and outwardly. The oil if it rectified, especially that which comes over first, is an excellent medicine against the Plague, Epilepsy, rising of the Mother and Megrim, 6, 8, 10, to 20 drops being taken thereof at once, and the nostrils also being anointed therewith for to smell it; and it is to be observed, that when it is rectified with spirit of salt, it proves much clearer, than done by itself without addition: but if it be rectified with salt of tartar, it is of much more virtue, though it fall not so clear, as that which is done by spirit of salt.

N. B. If it be rectified from a strong AQUA REGIA having before once already been rectified with spirit of salt, it will turn so subtle, that it is able to dissolve iron or copper in some sort, and to reduce them into good medicines; and in this second rectification by AQUA REGIA all will not come over, but part of it will be coagulated by the corrosive water, so that it turns thick, like unto mastich, which in the warmth is soft, and may be handled with ones fingers like wax, but in the cold it is so hard, that it may be broken and made into powder, and glitters like gold.

Of the Oil of Soot.

Of the soot, which is taken from Chimneys, where nothing is burnt but wood, there may be distilled a sharp volatile salt and a hot oil. The salt is in virtue not unlike unto that which is made of harts-horn or amber, and it quenches inflammation, from what cause so ever it does proceed: The oil may without rectification externally be used very successfully for all loathsome scabs, and for a scald head, etc. But if it be rectified, as has been taught to be done with the oil of Tartar, of Amber, and of Harts-horn; then it may safely be used inwardly, as the above written oil are used; for it does as well as these, yea better in some special cases.

How to Make a good Oil out of Soot without Distilling.

Boil the soot in common water, till the water turns blood red (urine is better than water) and set this solution (being in an earthen pot) in winter time into the greatest frost so long till all in the pot be frozen into one piece and turned white: then break the pot and the ice, and in the midst thereof you will find the hot oil unfrozen in color like blood, which is not much inferior in virtue unto that which is distilled, yet afterwards it may be rectified, and so exalted in its virtue, when you please, and it is to be noted, that this separation does only succeed in the greatest frost and cold, and not else.

Of the Spirit and Oil of Honey.

Of honey there may be made a subtle spirit and a sour vinegar, if it be mingled with twice as much of pure calcined sand and so distilled; and it falls much better yet if it be made with the flores of Antimony, which were taught to be made in the first part, whereby the spirit is increased in its virtue, and its running over hindered thereby; and so distilling it, there will come over a pleasant spirit, a sharp vinegar and some red oil also, which must be separated: the spirit after the rectification inwardly used is good in all diseases of the lungs. It opens and enlarges the Breast, strengthens the Heart, takes away all obstructions of the Liver and Spleen; it dissolves and expels the Stone, resists all putrefaction of the blood; preserves from, and cures the Plague; all Agues, Dropsies, and many other diseases, daily used from 1/24 to 1/8 ounce taken with distilled water proper for the diseases; the sour vinegar colors hair and nails as yellow as gold; it cures the itch and scabs of the skin; it cleanses and heals old and new wounds, they being bathed and washed therewith.

The red oil is too strong to be used of itself. It may be mingled with the subtle spirit which came over first and so used, and the spirit will be exalted thereby in its virtue.

Of the Oil and Spirit of Sugar.

In the same manner as has been taught of honey, there is also made a spirit and oil of sugar, viz. adding pure sand to it; or (which is better) of the flores of Antimony, and then according to the rules of Art one spoonful after the other of this mixture cast in, it will yield a yellow spirit, and a little red oil, which after the distillation must be digested in Balneo so long together, till the spirit have assumed the oil and be turned thereby very red in color; it needs not to be rectified, but may daily be used either by itself, or with such vehicles as are proper for your purpose; in all it is like in virtue unto that which was made of honey; yet this of sugar is more pleasant than the other; it renews and restores all the blood in man, in regard that it received great virtue from the diaphoretical flores of Antimony; and this spirit may fitly be used in all diseases, it can do no hurt, neither in cold nor hot diseases; it does help nature mightily, and does so much good, that it is almost beyond belief. Especially if for a time it be used daily from 1/24 to 1/8 ounce. The residue of it is black, and may be kept for the same use again, viz. for an addition to other honey or sugar, or else you may sublime it again into flores in the furnace described in the first part, or in the furnace described in the fourth part of this book, with an addition of iron or tartar, reduces it into a REGULAS, & etc.

To Distil an Excellent Spirit and a Blood Red Tincture of Corals and Sugar.

If you mix sugar with red corals made into powder and distil it, there will besides the spirit come over a blood-red tincture like a heavy oil, which is to be joined with the spirit by digestion in Balneo, and it will be as virtuous as that which was made with Antimony diaphoreticum. It does perfectly and lastingly cure the epilepsie in young and old; it cleanses the blood from all filth, so that the Leprosie together with its several species may be cured thereby, etc. Its use is the same as was taught above of the Antimonized spirit of sugar.

The Spirit of Must or New Wine.

Take sweet Must or juice of grapes, as soon as it is squeezed out, boil it to the consistency of honey, and then mix it with sand, corals, or (which is better) with flores of Antimony, and so distil it, and it will yield such another spirit as that which is made of honey or sugar, only that this is somewhat tarter than that of honey. With honey, sugar and the juice of grapes, several metals may be dissolved in boiling and so prepared and made up into divers medicaments, both with and without distillation, after the same manner as was taught above with tartar: for honey, sugar, and the juice of grapes, are nothing else but a sweet salt, which by fermentation and addition of some sour thing, may be changed into a sour Tartar, in all like unto that which is gathered in the wine vessels. There may be made also a tartar out of cherries, pears, apples, figs, and all other fruit, yielding a sweet juice; as also of rye, wheat, oats, barley, and the like, whereof in the third part more shall be said.

For every sweet liquor of vegetables, if it be turned inside out, by fermentation may be changed to a natural sour tartar; and it is utterly false, that (as some do suppose) only wine yields tartar, which by daily use made of it by those that have very hungry stomachs (like those of Wolves) indistinctly together with the nourishment went into the limbs, and there turned to a stony matter. If this were true, then in cold Countries, where no vine grows, men would not be troubled with the Gout or Stone; the Contrary whereof is seen daily; though I must confess, that among all vegetables none yields more than the vine, the concurrent acidity being the cause thereof; for it turns the sweetness into tartar; for the more sour the wine is, the more tartar it yields; and so much the sweeter, so much the less tartar. By this discourse an industrious Chymist may easily come to know the original nature and properties of tartar, and in default of wine, how to make it out of other vegetables; common salt or the salt of tartar may be distilled with honey, sugar, or sodden wine (SAPA) and it will yield such strong spirits, that metals may be dissolved with them, and they are not to be despised in Physick and Alchemy.

Oil of Olives.

Out of oils made by expression (as olive oil, rape oil, walnut oil, hemp-seed oil, linseed oil, and the like) there may be distilled a penetrating oil, useful both outwardly and inwardly, which is done thus: Take common potters clay not mingled with sand, frame little balls of it, as big as a pigeons or hens-egg, burn them (but not too strong) to a hard stone, so that they may attract the oil, and when they are no more quite red-hot, but pretty hot, then throw them into olive oil which is the best: let them lie in it, till they be quite full and drunk of the oil, which will be done in two or three hours (some cast them red hot into the oil, but amiss, because the oil contracts thence an EMPYREUMA) then take them out, and cast in one or two of them at once into your distilling vessel made red-hot, and let it go; and within a while after cast in one or two more, and continue this till you have oil enough. If the vessel be full of the balls, take them out with the tongs or ladle, that you may proceed without let in your distillation, and in this manner, you need not fear the breaking of your retort or receiver, or the burning of your oil. The distillation being performed, take off your receiver, pour the oil that came over into a glass retort, and rectify it from calcined Allome or Vitriol, and the Allome will keep back the blackness and stink, and so the oil will come over clear, which must be yet rectified once or twice more with fresh calcined Allome, according to the intentness of penetrating which you look for; that which comes over first, ought still to be caught by itself, and you will get a very fair, bright and clear oil, which is very subtle; but that which comes after is somewhat yellow, and not so penetrating as the first; and therefore it is but for external use to extract flores and herbs therewith, and to make precious balsoms for cold and moist sores. Also, you may dissolve with it Amber, Mastick, Myrrh, and the like attractive things, and with Wax and Colophony reduce it to a plaster, which will be very good in venomous sores and boils, for to attract the poison, and to heal them out of hand. If you dissolve in it common yellow sulphur made into powder, you will get a blood red balsom, healing all manner of scabs, and other like defects of the skin; especially when you add to it purified verdegrease, and in hot sores SACCHARUM SATURNI, which in a gentle heat and by continual stirring about do easily melt and mingle therewith. It needs not to be done in glasses, but may be done in an ordinary earthen pot or pipkin.

The Use of the Blessed Oil.

The first and clear is of a very penetrating nature: some drops thereof given in some AQUA VITAE, presently stays the collick, proceeding from winds that could not be vented; as also the rising of the mother, the navil being anointed therewith; and the cold humour being fain upon the nerves, whereby they are lamed; if you do, but anoint them with this oil, and rub it in with warm hands, it will quickly restore them, and therefore in regard of its present help, may well be called OLEUM SANCTUM. If you extract plates of iron or copper with this oil, it will turn deep red or green, and is a sovereign remedy for to warm and dry up all cold and watery sores. It consumes also all superfluous moisture in Wounds and ulcerous Sores, as also all other excrescencies of the skin: It heals tetters and scald-heads, and other like defects proceeding from superfluous cold and moisture. You may also dissolve in it Euphorbium and other hot gums, and use them against great frost, for what limb so ever is anointed therewith, no frost how great so ever can do it any hurt. The balsomes made with gum or sulphur may be also distilled by a retort, and in some cases, they are more useful than the undistilled balsome.

Of the Oil of Wax.

In the same manner, may be distilled also the oil of wax, the use thereof is in all like unto the former; and for all cold infirmities of the nerves, this is found more effectual yet than the former.

A Spirit Good for the Stone.

Out of stones which are found in grapes, there may be distilled a sour spirit, which is a certain and specifical remedy for the stone in the kidneys and bladder, and also for all pains of the gout. It is not only to be used internally, but also externally, wetting clothes in it, and applying them to the places affected, and it will assuage and drive away the pains.

Of the Spirit or Acid Oil of Sulphur.

To reduce sulphur into a sour spirit or oil has been sought hitherto by many, but found by few. Most of them made it in glass-bells, but got very little that way; for the glasses being quickly hot, could not hold the oil, so that it went away in smoke. Some thought to get it by distilling, others by dissolving, but none of all these would do the feat. Which is the reason why nowadays it is found almost no where right, and in the Drugstores and Apothecaries shops they usually sell oil of Vitriol instead of it, which by far is not to be compared in virtue to the oil of sulphur. For this is not only of a far pleasanter sour taste, but in efficacy also much exceeds the other. And therefore, being of so great use both in Physick and Alchemy, as in all hot diseases, mingling the patients drink therewith, till it gets a pleasant sour taste, for to quench the intolerable drought, to strengthen the stomach, to refresh the lungs and the liver. Also, externally to cure the gangrene. Also for to Crystallize some metals thereby, and to reduce them into pleasant vitriols, useful as well in Alchemy as Physick: I thought good to set down the preparation, though it be not done in this our distilling furnace, but in another way by kindling and burning it as follows.

    Make a little furnace with a grate, upon which a strong crucible must be fastened resting on two iron bars, and it is to be ordered so that the smoke be conveyed, not above by the crucible, but through a pipe at the side of the furnace: the crucible must be filled with sulphur even to the top; and by a coal fire without flame be brought to burn and kept burning. Over the burning sulphur, a vessel is to be applied of good stony earth like unto a flat dish with an high brim, wherein is always cold water to be kept, and whereunto the burning sulphur does flame: which thus burning, its fatness consumes, and the acid salt is freed and sublimed to the cold vessel, where it is dissolved by the air, and in the form of a sharp oil runs from the hollow vessel into the receiver, which must be taken off sometime, and more sulphur supplied instead of that which has been consumed, to the end that the sulphur may still burn in the crucible: and beat with the flame to the cold heat: and within few days you will get a great quantity of oil, which else by the (campagna) glass-bell in many weeks could not have been done.

N. B. Such a sour spirit or oil may also be got by distillation together with the flores, viz. thus: If you take pieces of sulphur as big as hens eggs, and cast them one after another into the hot distilling vessel, a sour oil together with flores, will come over into the receiver, which must with water be separated out of the flores, and the water abstracted from it again in a cucurbit, and in the bottom of your glass body you will find the oil, which in virtue and taste is equal to the former, but you get nothing near so much in quantity by this way, and if you do not look for the oil, you may leave it with the flores, which by reason of their pleasant taste of acid are much toothsome to take than the ordinary ones.

To the Courteous Reader.

Thus, I conclude this second part. I could have set down more medicinal processes in this Treatise: but having as many as will be a sufficient guide for the distilling of other things also, I thought it good here to acquiesce; and whatsoever has been here omitted, shall be supplied in the following parts.

FINIS.

THE COMPLETE WORKS OF RUDOLPH GLAUBER

trans: Chris. Packe

PHILOSOPHICAL FURNACES

THIRD PART

Third Part of Philosophical Furnaces

In which is described the Nature of the Third Furnace; by the help whereof, and that without Stills, and Caldrons, and other Copper, Iron, Tin, and leaden Instruments, various Vegetable burning Spirits, Extracts, Oils, Saks, & etc. by the help of certain little Copper Instruments, and wooden Vessels are made for Chymical and Medicinal Uses.

A Preface of the Copper Instrument and Furnace.

Now this Instrument is made of strong Copper plates after the following manner. You must make two strong hemispheres of Copper or Latten[15] of the bigness of a man's head (or thereabout) and join them together with a most strong solder, and that without tin, whereof the one must have a pipe. Now the pipe must be of a most exact roundness, that it may most accurately fit the hole that is made with an auger or wimble to keep the water from flowing out like to a tap, of the length of one span at least, wider on the hinder part towards the globe, than on the forepart, which also must be according to the bigness of the globe, greater or lesser, and be exactly joined with the best solder to its hemisphere, and the diameter of the forepart being very round like a tap, and most exactly filling the round hole must be of two fingers breath. Now there is required to the foresaid Instrument or globe, a certain peculiar little furnace made of Iron or copper, viz. most strong copper plates, covered within with stones or the best lute, into which is put this globe like a retort, so that it may lie

---

[15] An alloy of copper and zinc resembling brass. -PNW

upon two iron bars of the distance of a span, or span and a half from the grate; the neck whereof (that pipe) goes forth of the furnace one span at least. The furnace also must have below a place for the ashes, and above a cover with its hole for the letting forth of the smoke, and for regulating the fire, as you may see by the annexed figure. It must also below have a treefoot, on which the furnace must be set, and on the sides two handles by the help whereof it may be removed from place to place; the which is very necessary; for it is not only used for the distilling of burning spirits by wooden Vessels instead of copper, but also for such distillation, and digestion that is performed in gourds, bolt-heads, and other instruments of glass, stone, copper, tin, & etc. which are to be set in Balneo: also in the boiling of beer, metheglin, wine, and other drinks, which are to be performed by the help of wooden Vessels.

Of Wooden Instruments that are to be used instead of Stills, Baths, and Cauldrons.

In the first Figure, A represents the Furnace with a Copper Globe. B. The Copper Globe. C. The distilling Vessel. D. The Refrigeratory with a Worm. E. The Receiver. F. Stools on which the Vessels stand. The Second, A. Balneum with a Cover having Holes in it for the Glasses, set upon a Tree—foot. The Third, A Wooden Vessel for the making of Beer. The Fourth, A Pub for a making a moist bath, which is to be warmed by the Copper Globe. The Fifth, A wooden Box for a dry Bath to provoke sweat with Volatile Spirits.

In the first place, I shall speak of wooden Vessels that are to be used instead of Copper stills, in the distilling of burning spirits out of wine, beer, lees, malt, wheat, meal, roots, herbs, flowers, seeds, and other vegetables, as also oils of vegetables.

See that you have an oaken barrel, like to those wherein wine and beer are kept, of a just bigness, viz. answerable to the bigness of the globe, as is sufficient for the coction: for a barrel that is too big will make the coction slow, and tedious. A greater globe may be fitted to a lesser barrel, but not on the contrary, a great barrel to a little globe: For by how much the bigger the globe is, and the less the barrel, so much the sooner is the work hastened. Now seeing that this Art was invented for the saving of costs, which otherwise would have been expended in providing of stills, cauldrons, furnaces, & etc. it is best not to have too great a globe, which requires a greater furnace, and is more difficult to carry, because it is to be covered within with lute, or a wall; for it is sufficient if it be big enough for the coction. Therefore, I will give you a just and due proportion of both, viz. of the globe, and vessel, which in distillations and other operations, the courteous reader may Imitate.

A globe of the bigness of a man's head, containing three or four cannes, whereof each contains four pints, is sufficient for the heating of a barrel of 30, 40, 50, 60, and 100 gallons, which by how much the more remote from 100 and nearer to 30, so much the sooner is it heated, and the coction furthered; and on the contrary, by how much the nearer it is to 100 and more remote from 30, so much the slower is the coction. I do not therefore advise that a huge barrel be chosen for a small globe, because of a long and tedious operation. And if all and everything be not so accurately observed to a hair, yet it matters not much, because it suffices to do the same thing by the help of one small Copper instrument of divers forms. For in this way of distilling, wooden vessels that are requisite to the distilling of spirits, and boiling of Beer, and so for baths are more easily provided, then so many Copper vessels in the common way. For by this means not only costs are spared, but also it is instead of building of furnaces, because when any barrel has been used, you may remove it, and set another in the place of it for another operation, the which cannot be done with stills and cauldrons fastened into a furnace. And this invention is for those that want Artificers, as Coppersmiths, & etc. because wooden instruments are more easily provided: also by the help of this globe may most secret operations be performed.

For the furnace with the copper globe may be built in one place, and in another place the BALNEUM, viz. the places divided with a wall, so that he that looks to the fire may not know what is done in the Laboratory; for oftentimes the care of the fire is committed to heedless servants, that break glass instruments by their carelessness, by which means oftentimes a most precious medicine is lost; which danger this invention is without.

Wherefore this copper globe with its wooden vessels is more convenient than those copper stills and cauldrons. But this I would have you know: that this new invented distillation is slower, then the common way which is performed by stills, and consequently requires a long fire. I desire therefore the rich that dwell in large and spacious houses, that they would use the old way of distilling; but the poor, who have but little household conveniences, and the covetous, that they would use this little copper globe with its wooden vessels: for although there be a longer fire required, yet these are not to be compared to those costs which are otherwise expended upon so many copper vessels of so many divers forms. Let him therefore keep to his copper vessels, who cannot understand me, for it concerns not me. Without doubt, there are some whom this my new invented way of distilling will please, before other, being communicated for the sake of the poor laboring house-keepers, that cannot boil Beer, and distill burning spirits for lack of vessels: for a globe of five or four pounds is more easily provided, then other copper vessels, of 60, 80, 100 pounds: also, those wooden vessels are more easily provided than furnaces, which some for want of place only cannot build. Choose therefore which way you will, for these things which I have wrote, I have written for the poor's sake rather than for the rich. Certainly, rich men that have spacious Laboratories need not to be ashamed to follow this way, for it is free for every man to go a shorter way, unless they had rather prefer the old way before a new and compendious, whom I cannot help, being contented with a publication which is made for the sake of my neighbor, whether it be taken well or ill, with a good mind, certainly knowing that more profit then disprofit may be obtained by the help thereof. It shall not therefore repent him of his labor, who knows rightly to prepare and use this copper, and wooden vessels.

There follows now the Preparation of the Vessel.

The vessel being made is to be placed with one bottom, upon a stool that is fitted for it, which being done, make a hole with a wimble near the bottom, for the receiving of the neck of the copper globe, which is to be covered over with a linen cloth: make also about the lower bottom another hole for a tap, by the help whereof the remainder of the distillation is drawn forth: also you must make a large hole in the upper bottom, the diameter whereof must be one span for to pour in the water to be distilled, with a funnel. Also, there must be made a hole near the upper bottom of two or three fingers breadth, into which is to be put a copper pipe of a span long, which is to be fastened closely therein; and to this pipe another oaken vessel with a copper worm and cold water like to other refrigeratories, must be applied. Also, the joints of the aforesaid short pipe, viz. of the first barrel, and of the second barrel, viz. the refrigeratory must be straightly, and closely united together, which afterward may be the better joined together with a fit lute for the distilling. And this is the form and fashion of the wooden vessel, that is to be used in the place of copper vessels, in the distilling of burning spirits and oils. But you will object that these kinds of wooden vessels are porous, and drink up great part of the spirit and oils.

I answer, none of the spirits seeks a violent passage out, in case the ways be open. There is no danger therefore, when there is passage enough given them by a pipe that is wide enough. Neither does oil stick to them in distillation, for whatsoever is by force of the boiling water to be separated from the spices, and seeds that also is sublimable by the force of the seething water, to distil in, so that in the refrigeratory no more is lost than in the stills. Distillation being made, the aforesaid spirits may be rectified in these wooden vessels, (being first washed) as well as in copper stills.

The Making of a Wooden Vessel for a Balneum, which is to be used instead of Copper and Leaden Cauldrons for Digestion, and Distillation by Glass Vessels.

Make an oaken vessel as big as or as little as you please, according to the greatness, or littleness, multitude, or fewness of the vessels, of two or three spans high, a little narrower above then below, and so fashioned above, that a cover of wood, copper, or lead, may most closely be joined to it: the cover must have holes greater or lesser, according to the glasses, as is wont to be in the making of a BALNEUM, as you may see by the annexed figure. This vessel also must be placed upon a stool of the height of an ell, or such height as is required for the joining of the copper globe with the BALNEUM, which must have a hole near the lower bottom, for the receiving of the neck of the foresaid globe. In defect of such a vessel, which yet you may provide easily enough, take a wine or beer vessel divided in the middle, and make a hole near the bottom for the neck of the globe, make also a wooden cover with holes, & etc. He that will be curious may provide all things according to the best Art.

A Wooden Vessel serving for Boiling of Beer, Metheglin, Vinegar, & etc. as well as Copper, Iron, and Tin Vessels.

Make a wooden vessel, which shall be more high than broad, a little wider above than below, as you please: or take a wine or beer barrel divided in the middle, and near the bottom make a hole for the neck of the globe, which is to be covered with boards, which serves as well for the boiling of beer, & etc. as those of copper.

A Wooden Vessel for a Bath for Sweet, or Mineral Water, which may be according as you please, kept warm, for the Preserving of Health.

Make a long wooden tub convenient to sit in, which is to be set upon a stool of a just height, viz. that the bottom of the vessel may answer the neck of the globe which is put into the furnace: you may also have a cover, that may cover the whole tub, which may be divided and united in that place where the head goes forth, as appears by the annexed figure, or you may cover it with a cloth, laying it upon small crooked sticks fastened to the tub, yet so that the head may have its liberty, especially in a vaporous bath of common sweet, or medicinal water; or make a high wooden cover shutting very close, for a dry sweat, where it is no matter whether the head be shut in or no.

Of the Use of Wooden Vessels in Distilling, Boiling, Bathing, & etc. And, first of the Distilling Vessel.

He that will distil any burning spirit by help of the distilling vessel, out of wine, metheglin, beer, barley, wheat, meal, apples, pears, cherries, figs, & etc. also out of flowers seeds, and other vegetables, has need so to prepare his materials, that they may yield their spirit. Where I thought it convenient, and indeed necessary to say something of the preparation of each vegetable, for better information sake, or else a profitable distillation is not to be expected, but a labor in vain to be feared.

And first of the Preparation of the Lees of Wine, Beer, Hydromel, and other Drinks.

The lees of wine, beer, hydromel, & etc. have no need to be prepared, because they do easily enough of themselves yield their spirits, unless haply having lost all their humidity they be dried, which you may make moist again by the admixtion of common water, lest they be burnt in distilling & stick to the vessel; of which thing, more in the distillation itself. Now flowers, roots, herbs, seeds, fruits, apples, pears, cannot be distilled without a foregoing preparation. You must therefore first prepare them as follows.

Of the Preparation of all kinds of Corn, as Wheat, Oats, Barley, & etc. which must go before the Distilling of the Spirit.

And first of all, a malt must be made of the corn; as it is wont to be in the making of beer. Now the manner of making of malt is known almost to all, wherefore I need not speak much of that, because in all places that have no wine, there is scarce any house found in which Malt and Beer is not made, as well in the country as cities. But however, there is a great deal of difference of making of it, for a long knife does not make a good Cook, nor all drinkers of wine are good planters. For many have persuaded themselves that, if they follow the footsteps of their fathers, they have done well (although they have been in an error) and being scornful, refuse instruction. Wherefore something is to be said of the difference of malting. Although I never exercised the Art of making Beer, yet I am certain I do in that excel all other Distillers, and Brewers. For I often saw and indeed with admiration, the simplicity of many in their operations, although common, and daily, to whom though an age should be granted, yet they would never be thriftier, being content with their ancient customs. Good God! How perverse is the world, where no body labors to find out any good; neither is there any one that thinks of perfecting and amending things already found out: Where all things run to ruin, and all manner of vice increase: for now almost every one seeks only after riches BY RIGHT OR WRONG; for it is all one with them, if they have them, not thinking that things ill-gotten shall perish, and that the third heir shall not enjoy them, and that unjust riches shall devour those that have been honestly gotten, with danger also of eternal damnation. I pray you, if our Ancestors had been so negligent, and had left nothing to us: I pray you, I say, what Arts and Sciences should we have had now? It is come to this pass now, that virtues decrease, and vices increase.

Of the Difference of Malting.

The difference of malt, by reason whereof it yields better or worse beer, and spirit, consists for the most part in the preparation thereof; for being made after the vulgar way it retains its taste, wherefore it cannot yield good spirit, nor good beer, which is observed of very few, wherefore they could not draw forth good spirit out of corn, but such as favors of the taste and smell of the malt. Which is not the fault of the corn, but of the artificer not operating aright in the preparation. of his malt, in distilling and rectifying. For if it were prepared right in all things, corn yields a very good spirit, not unlike to that which is made from the lees of wine, in taste, odor, and other virtues. Which Art, although it be not known to all, yet it does not follow that it is impossible: Now I did not say that it is that common way, whereby that spirit, which is like to the spirit of wine, is distilled, but another which is more subtle, and witty. Out of all vegetables is drawn a burning spirit, yet such is perceived by some difference of the taste, and odor, but that is not the spirits faults, but of the vegetable, as of herbs, seeds, corn, & etc. communicating their virtues, taste, and odor to the spirit: whence that spirit deserves to be called not simple, but compounded, for else all the burning spirit (being rightly rectified from its flegme) is made out of anything, having the same virtues with the spirit of wine, although it seem improbable to some. I do not deny that one simple may yield more or less sweet spirit than another. For sweeter wines yield sweeter spirits: Also, clear wine yields a sweeter spirit than the lees of wine, although they come forth out of one and the same vessel. For clarified wine, and that which is separated from the faeces yields a sweeter spirit than the Lees, and impure and heterogeneal sediment, which corrupts the simple, and sweet spirit, with a strong taste and smell: so that that may deservedly, being as it were simple, be preferred before this which is accidentally corrupted. And this is to be understood of all other spirits. What has hitherto been said, has been spoken for the sakes of them, who have persuaded themselves that they could not perform chymical operations so well by the spirit of corn, as with the spirit of wine, for I never found any difference of them in the extraction of minerals and vegetables. Let him therefore that can, receive my opinion, and experience, seeing I will have nothing to do with contradicting Carper's.

Without hurt to others I dare not reveal the Art of distilling a sweet spirit with great profit out of corn, in all things like to that which made of the faeces of wine, viz. without the preparation or grinding of the malt, which shall haply be (see **Explicat. Mirac. Mundi**) communicated elsewhere at some time or other, for this Book is not written for the publishing of secrets, but of a new invented distillation. But you that will make a sweet burning spirit out of malt or honey; know this, that the corn must be brought after a certain peculiar manner into malt, and lose its ungrateful flavor before its distilling, and fermenting, or else after the wonted manner a certain ungrateful spirit will be drawn from thence, that cannot be compared to the spirit of wine. The whole Art thereof consists in a true preparation; for ungrateful things are by Art brought into gratefulness, and on the contrary grateful things are made ungrateful by negligence. And thus, much for information sake.

Of the Fermentation of Malt.

Take of Malt ground in a Mill as much as you please, upon which in a wooden vessel set upright, pour cold water, as much as will moisten it, and serve for mixtion and comminution; then also pour as much warm water as will suffice for the making the mixture moist and thin, and also warm; for it must be neither hot nor cold: which being done mix with it some new barm, and cover it with a cloth, and in a short times space, being exposed to heat, it will begin to ferment (wherefore the vessel is not to be filled to the top) and leave it so long in fermentation, until the mixture descends, which for the most part is wont to be done the third day, and the malt will be ready for distillation.

Of the Fermentation of Honey.

Neither has honey any need of a singular Art in its fermentation, because being mixed with 6, 7, 8, or 10 parts of warm water, it is dissolved, and unto the solution is added ferment, as has been spoken concerning malt, which afterward is left covered in some heat for to be fermented, being fit for distillation when it becomes to wax hot. Now know that too great a quantity of honey makes a very slow fermentation, viz. of some weeks and months; wherefore for acceleration sake, I advise that a greater quantity of water be added; although otherwise it yields plenty of spirits, but ungrateful, which therefore I advise nobody to distil as being unprofitable, unless anyone know how to take away the ungratefulness thereof. (See the ***Consolation of Navigators***.)

Of the Preparation of Fruits, Seeds, Flowers, Herbs, Roots, & etc.

The fruits of trees, as Cherries, Plums, Apples, Pears, Figs, Juniper—berries, Elder—berries, Dwarf—elder, and Mulberries, & etc. are bruised in wooden vessels, with wooden pistils; and upon them being bruised, is poured warm water, and ferment added to quicken it, as has been above said of malt. Seeds are broken in a mill, flowers, herbs and roots, are cut small, and are stirred up to fermentation by mixing of warm water, and barm or yeast.

An Annotation.

Before you distil the aforesaid vegetables prepared by the help of fermentation, diligently weigh, and accurately observe whether the mixture be sufficiently fermented, for sometimes there is too much cold, or hot water put to it; sometimes the vessel is not well covered, by which means the cold air is let in, whence the fermentation is hindered, and consequently the distillation of the spirit. For by the help of fermentation the burning spirit of the vegetables is set at liberty, without which it cannot be done; also the distillation is hindered by too much have, as well as by too much delay; for if you begin to distil before the time, viz. fermentation not being yet perfected, you shall have but few spirits; wherefore also the better part is, by many that are unskillful, cast to the swine, but without any great loss, if the matter were malt, because that swine are fed therewith; but not so if other vegetables were the matter of the distillation. Also too much slowness where the matter begins to be sour before it is distilled, yields very few spirits, that which often happens, whilst herbs, and flowers, & etc. are out of ignorance left in fermentation 3, 4, 5 and more weeks, before they be distilled, for the greatest part of the spirit is then turned to vinegar, which would not be so very ill done, if so be these men knew how to clarify the remainders, and turn it into vinegar, that nothing thereof might be lost; for the vinegars of herbs, flowers, seeds, and roots are not to be contemned. And so often times (a thing to be lamented) the better part, if they be spices, and precious things, is lost.

 The matter of the distillation, and other choice things, as seeds and herbs are cast away with loss; wherefore for admonition sake I was willing to add such things that the operators may have an opportunity to consider the matter a little more profoundly with themselves, or at least of learning the art of distilling from countrymen, who do not suffer their malt to putrefy, grow sour or mouldy, before they fall upon their distillations, but presently fermentation being made (the third or the fourth day) begin their distillation.

 But someone will object, that my vegetable Spirits are not pure because of the ferment that is mixed, having in itself a spirit, I answer, there is not so great a portion of the ferment mixed which can corrupt the vegetable spirit. For although some spoonful's of ferment yielding but a few drops of spirits be added to a great quantity of the vegetables; yet there can come no hurt or detriment to so many quarts of the vegetable spirit. I have seen some supercilious men that would not add ferment to the matter of their spirit, but sugar or honey, by which they would promote fermentation, and so have thought to get a pure spirit, not considering that honey and sugar, after fermentation are made to yield their spirit also, whereof one spoonful yields more than ten or twenty of Barm: But honey and sugar fermenting not without difficulty themselves, how can they promote the fermentation of other things? Who also have had experience, that the addition of their ferment has been superfluous, whilst their flowers and herbs have stood some weeks in maceration, before they begun to ferment, and that often they have contracted an acidity, mustiness and stink, the reason of which was an unsuitable ferment. There are indeed the fruits of some trees that have a sweet and full juice, as grapes, cherries, apples, pears, figs, & etc. which need not the addition of any ferment, having a natural ferment of their own, but other vegetables not so, being lean, as herbs, flowers, and roots. It is necessary there to promote the fermentation of them by the addition of a suitable ferment, lest in length of time these herbs and seeds lose their spirit exhaling in maceration. And this much I was willing to say for information sake, and indeed for the sake of them who seek after the best and choicest medicines, wanting a good burning spirit as a companion applicable to them. For this spirit came not only by itself, as AQUA VITA, into a medicinal use as well internal as external, especially that which is prepared of cordial, and cephalic herbs; but also, being united with the proper oils of those herbs in many desperate diseases, where it could put forth its virtues eminently.

    And this much suffices concerning the preparation of vegetables that goes before the distillation of burning spirits.

The Manner of Distilling in General Follows.

    He that is going to distil, has need to stir his fermented matter very well with a stick, that the thicker parts may be well mixed with the thinner, and then he must fill therewith his distilling vessel set upon a treefoot, and joined to the copper globe in the furnace on one side, and to the refrigeratory on the other, the joints in all places being well closed either with Oxen—bladders, or with starch and paper. Also, the interior part of the globe in the distilling vessel must be fenced with a copper or wooden basket, that the herbs, seeds, and other things enter not into the globe, into which only water must come. Also, the upper half must be close stopped with a fitting stopple wrapped about with linen cloths, (viz. that hole by which the matter to be distilled is put in) like to vessels of wine that are stopped. Which being well done, you must kindle the fire in the furnace under the globe, until all the matter in the whole vessel boil well, and that burning spirit rise, and go out, through the refrigeratory (where it is condensed) into the glass receiver that is set under it, no less than that distilled out of a still, and you must continue the fire till all the spirit be come forth, which you may know by the taste. Which being done, and all things being cold, let the remainders be taken out by the lower large tap—hole, for meat for swine, or other uses. The spirit that is drawn off may be exalted, and rectified at your pleasure in the same vessel, being first made clean together with the refrigeratory. Note well, that sometimes there is left a fat oil with the flegme in rectifying of the spirit, proceeding from that herb of which that was the spirit, which did distil off with the spirit from the matter with a strong fire in the first distillation, but in the rectifying, could not ascend with the spirit in a gentle fire, but is constrained to remain with the insipid flegme. And this oil also has its virtues, especially that which is rectified by a glass gourd in Balneo, with the spirit of salt, and clarified. Now the like oil is got almost from all herbs, roots, seeds, flowers, and fruits, but out of one subject more than another, according to the hot and cold temper thereof. Especially the sediment of wines yields a good quantity of such oil, which being rectified is a medicinal true oil of wine, but not before endued with a sweet savour, and it is an excellent cordial, although I know no body that knew this before.

And thus, I have showed the general way of distilling, burning spirits, by help of the aforesaid wooden distillatory. Now also follows.

The Manner of Distilling Spices, Seeds, Flowers, Herbs, Roots, Woods, & etc.

First, the seeds must be broken in a mill, flowers, herbs, and roots cut small, the woods broken or rasped, upon which afterwards a good quantity of water (in which they may swim) must be poured for the maceration of them, so that when the distillation is ended there may remain some water, lest for want of water they be burnt in the distilling, and yield an oil favoring of an EMPYREUMA, and not sweet. Neither is too great a quantity to be poured upon them, but as much as shall serve to prevent the burning of the aforesaid vegetables in the distilling of the oil thereof. And indeed, fresh vegetables may presently without any foregoing maceration, being put with their proper waters into the distilling vessel be distilled. But they be dry may for the apace of some days be macerated before they be distilled. Also, the water appointed for maceration must be salted, for the better mollifying, and opening the aforesaid materials, that they may sooner yield their oil. Now green and fresh need not any salt water, yet it will not be hurtful to mix some therewith, because salt helps the boiling water, to make the oil more easily to ascend. It also helps and furthers distillation as does Tartar and Allome, if they be rightly mixed and ordered. Which being all rightly done, the materials that are macerated must be put by a funnel into the distilling vessel, and fire must be given as has been spoken concerning the burning spirit, and the oil of the seed, or wood macerated in the water will come forth in the distillation together with the water. And although by this way more oil comes forth, viz. Maceration being made by the addition of salt, than without salt, by the help of the sweet water alone, as is the fashion in all places almost to distil oils of spices; yet much remains inseparable by the water, and consequently not to be sublimed with the water. Therefore, the better way is that which showed in the first part to be performed with the spirit of salt, which if you please you may follow. All the oil being come forth (which is perceived by the changing of the receivers) the fire is to be extinguished, and the remainder is to be taken out, which if it be of seeds, herbs, or fruits, may, being yet warm, be fermented by the addition of ferment for the distilling of the spirit, of which there cannot be so great a quantity by reason of taking away of the oil, as otherwise is drawn out of things that have not lost their oil: For all burning spirits partakes of much oil, of the essence, and nature whereof more a little after. Now spirits must be made without the addition of any salt, for salt hinders the fermentation, without which the burning spirit cannot be had. But the water that is distilled together with the oil, is to be set in a certain temperate place, until the oil ascends, and swim upon the water, from whence it is to be separated with a Tunnel (of which in the fifth part). Also, there are some oils which do not ascend, but fall to the bottom, which are also to be separated with a Tunnel, and kept for their uses. Now how these oils may be kept clear long, and not contract any clamminess, shall be taught in the fifth part: but how they may after they have lost their clearness by long standing, and are become tenacious, be restored and clarified again, is taught in the first part, wherefore I need not here repeat it.

How Oils are to be Coagulated into Balsomes.

It has been the custom a long time to turn aromatical oils into Balsoms, where always one has been willing to excel another in this Art, which nevertheless was nothing hitherto, but for a washing and cleansing; for they could not be used inwardly, but only outwardly for their odor to comfort the heart and brain. Now the aforesaid oils are coagulated many ways, and are made portable in Tin, Silver, and ivory boxes.

Some have mixed the fat of a lamb with them by help of heat, and have turned them into a liniment, which they have Colored with divers colors; as for example, they have corrupted the oils of green herbs; as rosemary, marjoram, lavender, rue, sage, with a green color, by the admixtion of virdigrease (which is noxious to the head and heart) where one corroborates and refreshes, another destroys. They have tinged the Balsom of Cinnamon, and LIGNUM RHODIUM with a red color by the help of a poisonous Cinnabar. Others that are more industrious, have tinged their Oils with extracted colors of vegetables, which balsoms are more safely taken inward: But are not durable, acquiring a sliminess and stink; wherefore they have mixed white wax to coagulate them: By which means they are become more durable without stinking; but yet at length of time so tenacious, that being smeared or rubbed upon the skin, they stick fast by reason of the wax that is mixed with them: at last others have found out a better way of coagulating aromatical oils, and other things, viz. by the addition of the oil of Nutmeg made by expression, having lost its odor and color by spirit of wine; which they called the MOTHER OF BALSOMS. And this way has been a long time concealed by Apothecaries as a great secret, until at length it is become common, so that balsomes prepared after this manner are sold in almost all shops.

But although that be the best way, yet they are not durable balsomes that are made that way, because they lack salt. I do not contemn and disapprove of Balsomes made after this way, for if a better way had been known, better had been made, for no man is obliged beyond his power. Wherefore they are not only to be excused, that have used Lambs fat, Wax, and the oil of Nutmeg in the making of their Balsomes, but also to be honored for their communication. Now seeing the aforesaid Balsomes cannot be taken inwardly, nor be so well outwardly administered due to their unctuosity, others have consulted to congeal the Oils by the admixtion of their own proper fix-salts: And Balsomes prepared after this manner are made free from clamminess, or tenaciousness, and may be dissolved in wine, beer, or any liquor.

Wherefore they may be not only conveniently taken inward, but also more fitly than those old, be rubbed outwardly for the odor's sake, because they are easily washed off again with water. They do not only give a most sweet odor being rubbed, but also because of the admixtion of the fixed salt, having the nature of salt of Tartar, do beautify the skin. Wherefore they are to be commended, being dissolved in fair warm water for a lotion for the head, and face; not only because they beautify, but corroborate with their excellent odor; which those fat Balsomes cannot do. Wherefore this way is to be preferred far before the other.

Let him therefore that will, receive what I have said, for RARE AND NEW THINGS ARE NOT ALWAYS ACCEPTED, especially BEING OBSCURE: but I hope for the approbation of the age to come.

The Manner of Preparing Follows.

Take the remains of the burning spirit, and being put into a sack, press it hard: reduce the water pressed out into vinegar, and of roses you shall have a rose vinegar, and of other things another, being the best in a Family for to season meats: then take the remains out of the sack, and reduce it to white ashes in a potters furnace, upon which pour the flegme of its own burning spirit (being separated) to extract the salt, from which evaporate again all the humidity in a glazed earthen pot: calcine the coagulated salt gently in a clean crucible, and it will be white and be like to salt of tartar in taste; from which abstract, sometimes its own proper burning spirit, calcining the salt first every time; and the spirit will be so exalted by its proper salt, that it will presently assume its proper oil, and will, being poured upon it, associate it to itself so as to be perceived no more in the spirit, which will remain very clear: Which being done, calcine the salt yet once more very well in a crucible, and dissolve so much of it in its proper flegme, as suffices for the coagulation of the oil, then mix this solution with the burning spirit, mixed with its oil, and set it in a vial of a long neck well stopped, in Balneo, that the spirit may not exhale, in the coction of it, and in the space of a few hours there will be an union of the mixture which will be as white as milk. Which being done, let the glass cool, for there is a conjunction of the spirit, oil, and salt, so that neither can be discerned from another, which is to be poured into a vessel of a wide mouth, and it will be congealed in the cold like a white ointment, not only to be anointed withal, but also to be dissolved in any liquor, being of an excellent odor, which may also be given inwardly very conveniently, and being used outwardly it makes the skin beautiful and sweet, wherefore this is that most desired balsome of Princes and Ladies. And by this way the three principles of vegetables, being separated, and purified, are again reunited, in which union there is found the whole virtue, taste, and odor of the vegetable.

Note well: he that will color balsomes, must draw the color out of vegetables with spirit of wine, which he must make to be coagulated together with it. After this aforesaid manner, therefore you may draw out of any vegetable that has in it salt, spirit and oil, soluble and well smelling balsomes without the addition of any other strange thing, which are not to be contemned.

    And because here also is taught that most odoriferous balsome of roses, for roses yield but a little oil, without which that cannot be done, know that only roses or rose leaves also are to be taken for the making the aforesaid balsome, but also together with the leaves those whole knots; for that yellow that is in them yields that oil, not the rose leaves, & etc. And let what has been said suffice concerning our preparation of balsomes, which if they be rightly made, are not I suppose, to be contemned, neither do I reject those that are made without salt: Let him that has better communicate them, and not carp at ours. And so I would that all and each process should be comprehended under some one general, viz. of distilling burning spirits, and oils, by the help of a wooden distilling vessel, and their conjunction by the help of their proper fixed salt, I could here add more things concerning the use, and virtues of spirits of wine, and of those most sweet vegetable oils; but because they are clearly enough spoken of by others, I account it a superfluous thing to repeat the sayings of others, being contented with the description of one only general process, which you may imitate in other particulars.

There follows now the use of the second wooden vessel, which is to be used instead of those of copper or lead, serving for distillations, digestions, extractions, and fixations.

The vessel being made ready according to the prescription set down before, there is nothing else to do, then to fit the furnace with the globe, and at your pleasure to heat water in it, with a government of the fire in the furnace. Now all things may here be done, which otherwise are done in a common BALNEO; where there is no other difference but of vessels; here is used a wooden vessel, there a copper, leaden, or iron, & etc. In this operation, also is used the same furnace with the same globe, which was used above in the distillation, wherefore you need add nothing else besides, for nothing is more common than a BALNEUM in distillation; let the demonstration therefore of the use of the copper globe suffice. Now I thought it worthwhile to set down some Chymical medicinal extracts, not common, which may be made by the help of this BALNEUM, which being rightly prepared do many things in many diseases.

And First a Vomitive Extract.

Take an ounce of the flowers of Antimony, of purified Tartar 2 ounces, of sugar—candy 6 ounces, of rain water two pints, being mixed together, set them in a strong vial in BALNEO for to be cocted, and make them boil strongly the space of ten or twelve hours. Then the BALNEUM being cold, take out the glass, and pour forth the decoction, and filter it through a brown paper put into a tunnel; the filtered water will be reddish betwixt sweet and sour, which take (the faeces in the Filter being cast away) and in a small gourd glass draw off all the moisture with a gentle fire in BALNEO unto the consistency of honey of a brownish color, upon which again pour a pint of spirit of wine, poured forth into a vial with a long neck; and set it in BALNEO with a moderate heat the space of eight or sixteen hours, and then the spirit of wine will separate, and extract the essence, which will be more pure and noble, the faeces being left in the bottom; which after all things are cold are to be separated by the help of Filtration through a double brown paper. Then take the red tincture that is filtered, and in a gourd glass in a gentle BALNEO draw off almost all the spirit of wine until there remain a matter like a very sweet syrup, which being taken out keep as a most excellent vomitive, most profitable in many diseases, where other Catharticks can do nothing. For this medicine works most gently, wherefore it may be given to children of a year and half old without danger, and also to old men. This medicine purges and attracts all humours from the nerves, and veins, opens all obstructions of the liver, spleen, lungs, and kidneys, by which means many most grievous diseases are cured.

    I never found a vomitive comparable to this, which works quickly and safely. The dose of it is from grain 1, 2, 3, 4, to 10, and 30 according to the age and sickness. It may be taken by itself, or in wine, beer, & etc. and it will within a quarter of an hour begin to work, and ceases within two hours. Sometimes it does not provoke vomit at all, but only stools, where a glyster is very helpful if it be given a little before the administering of the aforesaid medicine, being made of two or three spoonful's of oil Olive, and salt water; for the glyster prepares the way below, so that it seldom then works by way of vomit: when also the patient may presently after the taking of the medicine hold hot toasted bread to his mouth and nose, which hinders vomiting and promotes the operation by stool, but in my judgement it is better not to hinder the medicine seeking a spontaneous way of operation, and not forced: For vomiting is more convenient for some, than purging by stool. Now these things I have spoken for the sake of those, who although they abhor vomiting, yet desire to be purged by the essence of Antimony, which is of all that I know the most safe, and sweet Cathartick. For it searches the whole body far better than all others, and frees it from many occult diseases, the which all other vegetable Catharticks could not do. It has also this commodity in it, that although by littleness of the dose, or the strong nature of the patient it does not work by vomit or stool, yet it does not like other medicines hurt the body, but works either by sweat or urine, so that Antimony being rightly prepared is seldom administered without profit. Whereas on the contrary, vegetable Catharticks being given in less dose or because of some other causes do not work, although they do not make the body swell, and produce manifest diseases, yet they threaten to the body occult sicknesses.

    Now the ARCANUM of Antimony does not only not do hurt, if it does not sensibly operate, but by insensible working does much good to the body of man. Wherefore there is a great difference betwixt purging minerals, and vegetables. For minerals are given in a less dose without nauseousness, but vegetables with a great deal of nauseousness, and sometimes with danger to the sick in a greater dose. Now that nauseousness also proceeding oftentimes from the great dose of the ungrateful bitter potions does more hurt than the potion itself. I wish that such kind of gross medicines were abolished, and the sweet Extracts of Vegetables and Essences of Minerals were substituted in their place.

A Purging Extract.

Take of the roots of black Hellebore gathered in fit time, and dried in the air, one pound, the roots of Mechoacan, Jallap, of each four ounces; Cinnamon, Anniseed, and Fennel—seed, of each one ounce; of English Saffron a dram, powder all these ingredients, then pour upon them the best rectified spirit of wine, in a high glass gourd, and upon this put a blind Alembick, and set it in digestion in Balneo until the spirit of wine be tinged red, which then decant off. Then pour on fresh, and set again in digestion; until the spirit be red, then pour on fresh again, and do this so often until the spirit will no more be tinged red, which commonly is done at three times. Mix these tinged spirits, filter them, and in Balneo by a glass Alembic, with a gentle heat draw them off from the Tincture, and a thick juice will remain at the bottom of a brownish color, which you must take out whilst it is yet hot, and keep it in a clean glass for its uses. The Spirit of Wine drawn off from the extract may be reserved for the same use. Now this extract is given from grains 3, 6, 9, 12, to 31; according to the age and strength, being mixed with Sugar, it has not an ungrateful taste, and it works gently, and safely, if it be not given in too great a dose. And if you will have it in the form of a Pill, while it is still hot mix with it an ounce of clear Aloes, and half an ounce of Diagridium powdered, being mixed bring it into a mass for Pills, and keep it for your use. The dose is from grain 1, to a scruple. It evacuates all superfluous humours, but it is not to be compared with the medicine of Antimony. And this extract I put down for the sake of those that fear Minerals, and abhor Vomits, which in my judgement is the best of all vegetable Catharticks.

A Diaphoretical Extract.

Take the wood of Sassafras, Sarsaparilla, of each six ounces; Ginger, Galengal, Zedorary, of each three ounces; long Pepper, Cardamoms, Cubebs, of each an ounce; Cinnamon, Mace, of each half an ounce; English Saffron, Nutmeg, Cloves, of each a dram: Let the woods be rasped, the roots and spices powdered, pour upon them, being mixed, the spirit of wine, and let the tincture be drawn forth in Balneo, as has been above said of the purging extract. Evaporate away the spirit to the consistency of honey, which keep for your use. It is good in the Plague, Fevers, Scorbute, Leprosie, French-pox, and other diseases proceeding from the impurity of the blood, curing them by sweat. The dose of this Extract is from a scruple to a dram with proper vehicles: it provokes sweat presently, drives away all venenosities from the heart, and mundifies the blood.

And although it be a most effectual vegetable Diaphoretick yet it may not be compared to those subtle spirits of minerals, of which in the second part. Also, animal diaphoreticks have their commendations, as the flesh of vipers, the fixed salt of spiders and toads, in their peculiar operations, where each alone without the mixture of any other thing puts forth and shows its operations;

neither are animal and vegetable diaphoreticks to be compared to the mineral, as BEZOARTICUM MINERALE, ANTIMONIUM DIAPHORETICUM, and AURUM DIAPHORETICUM.

A Diuretical Extract.

Take the seeds of Saxifrage, Carroway, Fennel, Parsley, Nettles, of each 3 ounces, the root of liquorish, the greater burr, of each an ounce, the powder of woodlice half an ounce. Let these being mixed and powdered be extracted with spirit of Juniper according to art: then mix these following things with the extracted matter: Take the salt of Amber, Soot, Nettles, of each half a dram, purified Nitre a dram. Let these be powdered, and mixed with the extract and this mixture be kept for use. The dose is from a Scruple to a dram, in the water of parsely, fennel, & etc. This extract forces urine, opens the ureters, purges the reins and bladder from all viscous flegme (the mother of all tartareous coagulation) viz. if it be used timely. In this case is commended also the solution of flints, and crystals, made with spirit of salt. A greater commendation has salts of nephritick herbs made by expression, and crystallization, without calcination, the preparation whereof shall not here, but elsewhere be taught.

A Somniferous Extract.

Take of THEBAIC OPIUM four ounces, of Spirit of Salt two ounces, purified Tartar one ounce, set them being mixed in maceration in Balneo in a glass vessel for a day and night, and the spirit of salt with Tartar will open the body of the OPIUM, and prepare it for extraction, upon which pour half a pint of the best spirit of wine, set it in a gentle Balneo to be extracted. Decant off the spirit that is tinged, and pour on fresh, set it in digestion till the spirit be colored. Then mix the extractions together, and put to them in a fresh glass gourd two drams of the best Saffron, of oil of Cloves a dram, and draw off the spirit of wine in Balneo, and there will remain a thick black juice, which is to be taken out, and kept in a clean glass vessel. The dose thereof is from one grain, to five or six, for those of a man's age, but to children the sixth or eight part of a grain. It may be used in all hot distempers without danger. It provokes quiet sleep, mitigates pains as well outward as inward, it causes sweat; but especially it is a sure remedy for the epilepsie in children that are new-born; for as soon as it is given to them to the quantity of the eighth part of a grain in wine, or woman's milk, there presently follows rest, and sweat with sleep, by which means the malignity is expelled, the children are refreshed, and desire victuals, and the fit returns no more afterwards. Although haply the like symptoms may be perceived again, yet if the aforesaid dose be administered again, the children are refreshed, and cured wholly, whereas otherwise they would have died, & etc. whereof I have not restored few with this medicine. Moreover, also there are very effectual anodyne medicines, as those volatile spirits of vitriol, allome, Antimony, and other minerals, with which, as also with that narcotick sulphur precipitated from the volatile spirit of vitriol, nothing may be compared.

A Cordial Extract.

Take red roses four ounces, of the lily of the valley two ounces, the flowers of borage, rosemary, sage, of each an ounce; cinnamon, lignum aloes, of each two drams; cloves, mace, nutmeg, galangal, cardamoms the lesser, of each half an ounce; the shavings of ivory, harts-horn, of each an ounce; of ENGLISH saffron a dram, of NUX-VOMICA a dram. Mix them and reduce them to a fine powder, and let the tincture be extracted with spirit of wine in Balneo, which is to be drawn off again, unto a just consistence. Let the extract be kept for use. It may be used in almost all faintings, and other affects that are not joined with a preternatural heat. The dose thereof is from grains 3, 6, 9 to a scruple with proper vehicles; being often administered it refreshes the spirit, corroborates the brain, and other parts of the body. It is made more efficacious by the adding of the essences of minerals, especially of gold, of which thing see the first part concerning the sweet oil of gold.

Of an Odoriferous Extract.

I need not teach the making of any odoriferous vegetable extract, because the manner of drawing forth, or distilling oils of vegetables that have sweet odors has been shown a little before, as of herbs, flowers, and seeds, which are the most noble, and sweet essences of vegetables, by the odor whereof the heart and brain are corroborated, which being reduced into balsomes are made transportable. Better extracts, therefore, and more excellent cannot in my judgement be made out of vegetables, than those aforesaid oils, unless any one would mix aromatical extracts made with spirit of wine with metallick solutions, and being mixed digest them, then there will certain most odoriferous oil go from the extract not only more efficacious, but more excellent than that common distilled oil by reason of the admixtion of the spiritual metallick virtue, especially of gold and silver, dissolved in the acid MENSTRUUM communicating its virtues to the Aromatical oil. Moreover, any vegetable oil may be exalted in virtues and odor by the help of spirit of urine, or salt Armoniack, by the help whereof not only odoriferous oils are exalted, but also the inodorous oils of vegetables are made odoriferous, if they be a while digested in spirit of urine: and not this only but every mineral, and metallick sulphur, although the odor thereof be bound up with most strong bonds, is opened by the benefit thereof, and is reduced by digestion in a very little time into a most sweet and odoriferous essence. Lixivial spirits exalt the odors, and colors of sulphurs; acid purge sulphurs, but change their colors and odors. Musk and Civit get the sweetness and excellency of their odor from subtle urinous spirit of a certain Cat, digesting some certain fat and converting it into such a kind of most odoriferous matter.

    And let this that has been said suffice concerning Extracts, which might have been omitted, because many of these kinds of Extracts are found in the writings of other authors in many languages: But I was willing to set down these, lest this book might seem to contain in it nothing else besides the new way of distilling, being furnished also with good medicines.

Of Baths.

A little before has been given a description of a Tub for a Bath in which any one may sit with his whole body except his head, not only to be washed in sweet warm water, whether medicinal and mineral, but also to sweat in without water, where the vessel is heated by warm vapors, either of sweet waters, or minerals. And every one may provide such Baths for himself according to his necessity at home, whereby the same diseases are cured as those that are cured by the help of natural Baths, so that he need not for the baths sake go a great journey, but may stay at home with his family and follow his Calling without trouble, when he has occasion and need to use them.

And whereas it cannot be denied, that using the Baths most grievous diseases which cannot be cured by Physicians, are happily cured; I was willing for the sake of my neighbor to publish this instrument together with the preparation of mineral waters; which publishing will not without doubt be without profit, and advantage. Wherefore I will in brief show you the preparation of mineral, and sweet waters, and their use, and first:

Of a Bath of Sweet or Common Water.

There is no art to make a Bath of sweet water, for you have nothing else to do then to fill your vessel with river or rain water, and to make a fire, which by the help of the copper globe will heat the water, which being sufficiently heated, you may sit in it, and cover the Tub, that the hot vapors evaporate not, nor the cold air enter in, and cool the exterior parts of the body. Wherefore also you must apply a clean linen cloth about your neck, lest the warm vapors evaporate there: which being rightly observed, you may sit the space of 1, 2, 3, hours, or as long as you please or your sickness may require. You must keep a continual heat as much as is necessary, which may be done by the help of that globe. If you be thirsty in the mean time you may drink some proper distilled drink according to the nature of your disease, of which thing nothing now, because I am resolved to write a peculiar BOOK DE BALNEIS, and here only show the use of that copper globe in heating of Baths. And although there be not a perfect instruction of all, yet of some Baths, and their uses there shall a short instruction be given in this place.

Of the Nature and Property of Natural Baths.

Know that the greatest part of medicinal waters in GERMANY, and other countries, as well hot as cold carry with them from the earth a certain sulphureous acidity, more or less, in which acidity consists that medicinal faculty and virtue of this or that water. And if those waters lose their odor and taste by the exhaling of their subtle spirits, then also they lose their virtues; although also there be found some waters, which have not only a spiritual sulphur, but also are impregnated with a certain mineral, or metallick body mixed with Allome, or Vitriol, which comes not elsewhere then from the common water running through the mines. There are found also other baths, the power and virtue whereof consists not in any spiritual sulphur, nor in any metallick body mixed with salt, but only in a certain spiritual salt mixed with a certain subtle fixed earth, which waters do not run through metallick mines as others do, but rather stones of the mountains calcined with a subterranean fire, whence also they borrow their subtle acidity with their insipid earth. And this no man will deny that has a knowledge of volatile and fixed salts of minerals, and metals: the which I am able to demonstrate with very many, and most evident reasons, if time and occasion would permit; but it shall be done sometime or other as has been said in a particular treatise. Now therefore I will only teach how by salts, minerals, and metals, artificial Baths may be made, which are not only inferiour[16] to the natural in virtue, but also oftentimes far better, and that without much cost or labor, which any one may use at home instead of the natural for the expelling of diseases, and recovering of health. And although I am resolved to set forth a book that shall treat largely of the nature, and original of Baths, and of their use; yet I am willing now also to say something in brief concerning it, and that from the foundation, seeing that there are so many different opinions of learned men, and those for the most part uncertain.

As concerning therefore, the original of the acidity as well volatile, as corporeal, as also the heat of baths, know that is not one, and the same; for else each would have the same properties, but daily experience testifies the contrary: For it is manifest that some Baths help some diseases, and others are hurtful for them, which comes from nothing else but from the difference of the properties of the mineral waters proceeding from a diversity of mines impregnating those waters. In a

---

[16] I compared this word with that in the original printed edition of 1689, page 64. The original word is indeed "inferiour", although I believe it is a typesetting error. - PNW

word, sweet waters attract their powers, and virtues in the caverns of mountains from some metal and minerals of divers kinds, that have naturally a most acid spirit of salt, as are divers kinds of marcasites containing copper and iron, and sometimes gold and silver; also kinds of vitriol and Allome called by the ancients MISII, RARII, CHALCITIS, MELANTERIA, and PYRITIS, whereof some are found white like metals, but others dispersed in a fat earth, of a round figure in greater or lesser pieces: which sulphureous salt mines while the water runs through, and humectates, that spirit of salt is stirred up, having got a VEHICULUM, and falls upon the mines by dissolving them, in which solution the water waxes warm, as if it had been poured on quick lime, or like spirit of vitriol, or salt mixed with water, and poured on iron, and other metals; where continually and daily that water running through the mines whose nature and properties it imitates, carries something with it: wherefore there are so many, and such various kinds of Baths as are the mines by which the water is heated. Let him that will not believe take any mineral of the aforesaid quality, and wrap it up in a wet linen cloth for a little while, and he will see it experimentally that the mineral stone will be heated by the water, and so heated, as if it were in the fire, so as you can scarce hold it in your hand, which at length also by longer action will cleave in sunder and be consumed like quick lime.

    I will publish some time or other (God willing) more fully, and clearly in a peculiar treatise this my opinion, which I have now delivered in very few words. Although to the sick it be all one, and it matters not to them, from what cause the baths come, and whence they borrow their virtues, if so be they may use them; this controversy being left to natural Philosophers that will controvert it, which none of them can better decide than a skillful Chymist, that has the knowledge of minerals, metals, and salts.

And First of sulphureous Baths that have a subtle Acidity.

In the second Treatise, I have demonstrated the manner of distilling subtle, volatile, sulphureous spirits, viz. of common salt, vitriol, allome, nitre, sulphur, antimony, and other salts of minerals, and metals, and their virtues, and intrinsical properties, now also I will show their extrinsical use, as they are to be mixed with waters for Baths. The virtues therefore of Baths proceed not from insipid water, but from those most subtle, volatile, sulphureous, and salt spirits; but these being of themselves not mixed with water unfit for Baths, to be used for recovering of health, by reason of their too great heat, and subtlety; the most high God has revealed to us unworthy and ungrateful men his fatherly providence showing to us by nature the use of them, and the manner of using of them for the taking away of diseases; which (nature) being never idle, works incessantly, and like a handmaid executes the will of God, by showing to us the various kinds of distillations, transmutations, and generations. From which teacher we must learn all arts and sciences, seeking a certain, and infallible information, as it were out of a book writ with a divine hand, and filled with innumerable wonders, and secrets. And this is a far more certain knowledge then that empty, and imaginary Philosophy of those vulgar disputing Philosophers. Do you think that that true Philosophy can be sold for a hundred Royals? How can anyone judge of things hid in the earth, who is willfully blind in things exposed to the light of the Sun, hating knowledge? I wish knowledge were suitable to the name: how can anyone that is ignorant of the nature of fire, know how to work by fire? Fire discovers many things, in which you may as in a glass see things that are hid: The fire shows to us how everything, waters, salt, minerals, and metals, together with other innumerable things are generated in the bowels of the earth by the reflexion of that central, and astral fire: for without the knowledge of fire all nature remains veiled, and occult. Fire (always held in great esteem by Philosophers) is the key for the unlocking of the greatest secrets, and to speak in a word, he that is ignorant of fire is ignorant of nature with her fruits, and he has nothing, but what he has read, or heard, which oftentimes is false, according to that: HE EASILY SPEAKS UNTRUTHS THAT SPEAKS WHAT HE HAS HEARD.

He that is ignorant knows not how to discern betwixt the truth and falsehood, but takes the one for the other. I pray you, you that are so credulous, do you think that your teacher wrote his books from experience, or from reading other Authors? May they not be corrupted and sophisticated by antiquity, and frequent description? Also, do you understand the true, and genuine sense of them? It is better to know, than to think: for many are seduced by opinions, and many are deceived by faith that is without knowledge.

And thus, much for youth's sake I was willing to say, that they would not spend their tender years in vanities, but rather would make trial in the fire, without which no man obtains a true knowledge of natural things; which although it seem hard in the beginning, yet it is pleasant in old age.

Now follows the mixture of those Subtle mineral, sulphureous, and salt spirits with water.

As concerning the weight of the aforesaid spirits that are to be mixed with sweet water, giving it the nature, and property of natural baths, I would have you know, that of those, which in the second part I showed to be various, and divers, being, viz. not equal in virtue, the same weight cannot always be so accurately observed: seeing also there is a Consideration to be had of their strength, and of the strength of the patient.

Now you may at the beginning mix one or two pounds of the spirits with a sufficient quantity of the water, and then by sitting in it make trial of the strength of the artificial Bath, which if it be too weak is to be increased by adding a greater quantity of the spirits, but if too strong, then it is to be diminished by abstraction; of which more at large in ARTE NOSTRA BALNEATORIA. Now this observe, that it is best to make Baths in the beginning weak, then stronger by little and little by degrees, as the nature of the sick is accustomed to them, that it be not overcome by the unaccustomed use of them being too strong. Wherefore Baths are to be used with discretion, and cautiously, for which matter I refer the reader to my APTEM BALNEATORIAM, in which he shall find plain, and perfect instruction; let it suffice therefore that I have showed the use of the Copper Globe, in heating Baths, which let the sick take in good part, until more come. Now follows the use.

Of Sulphur Baths.

Apply the furnace with the Copper Globe to the Tub after the manner aforesaid, and pour in a sufficient quantity of sweet water, which make hot with the fire kindled in the furnace by the help of a globe: which being sufficiently warmed make the patient sit in it, and pour into it so much of the sulphureous spirit as is sufficient; which being done cause that the tub be covered all over, that the volatile spirit vanish not, and as necessary requires, continue the heat till the patient come forth. Know also that the water is to be changed every time, and fresh spirits to be mixed. And this is the use of the Copper globe, in heating baths of sweet or medicinal water, and that either of vegetables, or mineral, and this made sulphureous by art or nature; whereby most grievous, and otherwise incurable diseases are happily cured: Of which enough now in this Treatise.

The use of the Copper Globe in Dry Baths, which are more excellent than the moist in many cases.

I might have put off this matter into its proper Treatise, where all things shall be handled more largely, and clearly: yet by reason of some unthought of impediments for a while procrastinating the edition of the promised Treatise, I am resolved to say something of their use after I have made mention of the humid, and indeed not only of the use of those subtle, sulphureous, and dry spirits, but also of the use of subtle, vegetable and animal spirits which are medicinal, because in some diseases dry baths are more commodiously used, than moist. He therefore that will provoke sweat by a dry bath without water, let him provide a wooden box, or wooden instrument convenient to sit in, standing upon a stool bored through that you may raise it up, more or less, according as you please, and having boards appointed for the arms and feet to rest upon. This box also besides the great door must have also a little door serving for the putting in of a burning lamp with spirit of wine, or of any earthen vessel with coals for to heat it. (See the sixth figure.[17]) The box being well warmed, let the patient go in, and sit upon the stool, let the box be very close shut all about, and the furnace with the Copper Globe be fitted thereunto, under which let there be a small fire kindled, by help whereof the volatile spirit growing warm, goes forth into the box like a most subtle vapor, penetrating all about the patient. But when this spirit is not sufficient to heat the box, set in it a burning lamp with spirit of wine, or some earthen pot with coals (the best whereof are made of Juniper or the vine, especially of the roots as being such that will endure long, and cannot easily be extinguished by the vapors of those spirits) that the patient take not cold, and the vapors of the spirits may the better penetrate the body of the patient. Let the wick for the spirit of wine in the burning lamp be incombustible made of the subtle threads of gold, of which thing more in ARTE BALNEATORIA. In the meantime, that volatile spirit penetrates, and heats the whole body, and performs its office, being this way used better than by being mixed with water. When the patient has sat there long enough let him come forth, and go into a warm bed to sweat. Now before he goes into the box let him take a dose of that

---

[17] The bottom right figure in the engraving near page 250, at the beginning of Third Part of this book. -PNW

volatile spirit, inwardly which is used outwardly to provoke sweat, and accelerate the action. And by this means not only those volatile sulphureous spirits of salts, minerals and metals, are used outwardly without water to procure sweat, but also the spirits of many vegetables, as of mustard seed, garden cresses, crude tartar, also of animals, as harts—horn, urine, salt Armoniack, & etc. for the expelling of most grievous, and desperate diseases. Now the aforesaid spirits have diverse properties, the volatile spirits of salt, minerals, and metals have some, those of vegetables and animals have others; those have a sulphureous and fiery essence; these a mercurial, and aerial; wherefore they serve for different uses. In some diseases those sulphureous are preferred, but in others vegetable and animal, where also a consideration is to be had of the sickness, and has itself, that one be not used for the other, to the great damage of the sick. For almost all natural baths, and volatile spirits of salts, minerals, and metals, partake of some most subtle, penetrating, heating, and drying sulphureous salt spirit; but the spirits of vegetables, and animals partake of a certain volatility that is most subtle, penetrating, heating, opening, cutting and attenuating, both urinous and nitrous, viz. contrary to the former; as appears by the pouring on of any volatile sulphureous spirit, as of common salt, vitriol, allome, minerals, and metals, upon the rectified spirit of Urine, or salt Armoniack: where presently the one mortifies the other, and takes away its volatility, and subtlety: so that of both subtle spirits of divers natures there comes a certain salt of no odor and efficacy. Whence it is manifest that all spirits partaking of divers natures, and essences have not the same faculties.

Therefore, be cautious in giving most potent spirits, lest you give an enemy instead of a friend, and learn their natures, virtues, and essences, before you use them in medicine. But you ask, whether is that great force of those spirits gone as it were in a moment? Did it evaporate in that duel? No, I say, but transmuted into a corporeal substance, for of a most pure, mineral, subtle, and most volatile sulphur, and a most penetrating animal MERCURY is made a certain corporeal salt, which is wonderful, and deserves to be called AQUILA PHILOSOPHORUM, because it is easily sublimed with a gentle heat, in which many things lie: for it does not only conduce to the solution of metals, especially of gold, but also of itself by the power of maturation does become a most efficacious medicine: Of which no more at this time, because I will not only advise the reader, that he be diligent in searching out the nature of spirits, which although they change their bodies, yet are not therefore to be called dead, but rather reduced to a better perfection. And let this suffice concerning the dry use of baths in provoking sweat for the expelling of diseases: now for what diseases this or that spirit serves, you shall find in its proper Treatise, of which there has been mention above, but in a word, know that those volatile sulphureous spirits of salts, minerals, and metals, are good in all obstructions of the inward parts, viz. of the spleen, lungs, and liver, but especially are most excellent in heating the cold nerves, because they do most efficaciously heat, attenuate, cut, expel, and mundify, wherefore they are good in Contractures, Palsies, Epilepsy, Scurvy, Hypochondrical Melancholy, Morbus Gallicus, Itch, and other corrosive ulcers, and Fistulaes, & etc.

But the spirits of another kind, as of Tartar, Harts-horn, salt Armoniack, Urine, & etc. are hot also, but not so dry, and besides the heating virtue, have also a penetrating, cutting, mollifying, attenuating, absterging, and expelling power; wherefore also they work wonderfully in all obstructions of the inward and outward parts: for they do better than all others, open the pores of the skin, and provoke sweat, mollify, and open the hemorrhoids; provoke the MENSES of young and elder women, purge and heat the womb, and therefore cause fruitfulness; they heat and purge a cold and moist brain, actuate the intellect, and memory, let they that be great with child take heed of them, and also they that have a Porous open skin. Such and other more properties, and that deservedly are ascribed to these spirits. Now those two aforesaid baths (in one whereof those spirits are used in a humid way, being mixed with warm water, for the whole body to be bathed, and sweat in, but in the other in a dry way where the vapors are by force of the fire made under the Globe, forced up into the sweating box towards the patient, which being used after this manner do oftentimes penetrate, and operate more efficaciously than the humid way) are not to be slighted for the recovery of health, as doing things incredible. Now those spirits not being found in shops, nor being made by any according to the manner that I have showed in the second Part, I would have you know that there is yet another matter, which needs not to be distilled, and it is mineral; which being put into the Copper Instrument, does of its own accord without fire yield such a sulphureous spirit, which penetrates very much, and goes into the sweating box, like in all things to that which is made out of salts, minerals, and metals. Nature also has provided us another matter that is to be found everywhere, which being in like manner put into the Instrument does by itself, and of its own accord without fire yield a spirit, in virtue not unlike to that which is made from crude Tartar, or salt Armoniack, Soot, Urine, & etc. Of which in the second Part, doing, viz. the same things with that which is made with costs and labors. Those foresaid two matters therefore can do the same things, which are required for a bath and sweating, which those two foresaid kinds of spirits, viz. mineral and sulphureous, vegetable and animal can do, & etc. Now what those two matters which are easily everywhere to be found are, you desire to know; but I dare not if I would, for the sake of the pious to reveal them because of the ungrateful, and unworthy. For it is an offence to cast pearl before swine, which yet the pious may, by the blessing of God find out by the reading of the rest of my Writings.

Now follows a wooden vessel which is to be used instead of a Cauldron in boiling of Beer, Metheglin, Vinegar, & etc.

Many things might be said concerning this matter, for although men may be found in any part of the world, who know how to make malt of corn, and of this beer and vinegar; yet many things may be said of this matter for the correcting of it; but because it is not my purpose to show such things now, yet I shall say something of the use of the copper globe which any one may provide instead of Caldron, and which is to be used with a certain wooden vessel in the boiling of beer, which by this way he may, as has been spoken above concerning the operations, make as well as by the help of Caldrons. Moreover I

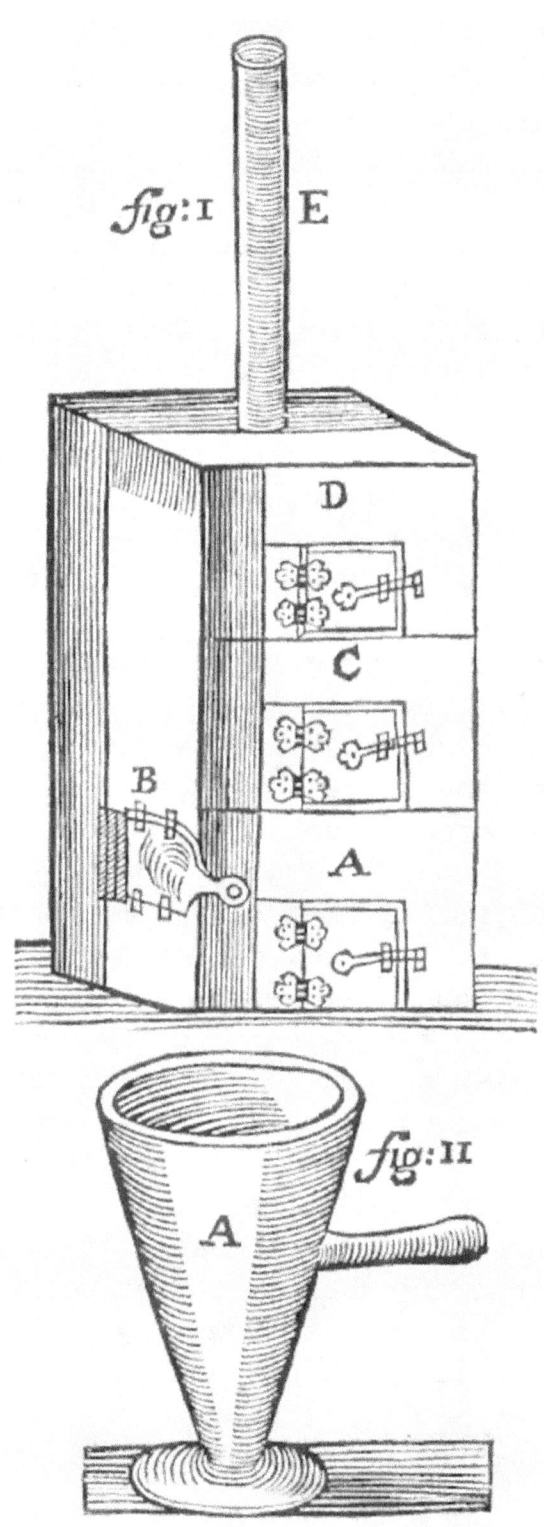

could here also teach some other most profitable secrets, viz. (See **Consola. of Seamen**) how honey may be freed from its ungrateful odor and taste by the help of precipitating; and how afterwards a most sweet spirit is to be drawn out of it very like in all things to the spirit of wine: also how the best and sweetest wine clear, and durable like to Mallago, may be made thence: also how after purging it is to be crystallized, so as to resemble Sugar-candy in goodness and taste: also how the sweetness thereof may be converted into Tartar, very like to the natural: (See **Explicat. Mirac. Mundi.**) Also how out of fruits of trees, as cherries, apples, pears, & etc. a very good, and durable wine in goodness, color, taste, and virtue, like to the natural, may be made; also how out of unripe grapes, that are not matured either by the inclemency of the country, or air, their acidity being changed into sweetness, a very good Wine like to the Rhenish may be made: also how out of sorrel, and other vegetables, a very good Tartar may be made, and that in a great quantity without much cost, resembling the Rhenish in color, taste, and other virtues: Also how out of Corn, (whether malted, or ground in a Mill) a very good spirit is to be made, and also a very good vinegar like to the Rhenish; also how out of Corn, (ground in a mill) or Meal, a very good spirit is to be distilled without any loss of the Meal, continuing yet fit to make bread. Such and more of this sort might be taught in this place, but because it is not good to divulge all things together, and at once, and this book would by this means grow bigger than I am willing it should. If such things should be here taught, I shall make an end of this Book (omitting other excellent possible secrets of nature) which although it be but little, yet will without doubt be profitable to many. And so, Reader farewell.

Fourth Part of Philosophical Furnaces

In which is described the Nature of the Fourth Furnace; by the help whereof, Minerals and Metals are tried, and examined after a more compendious way, than hitherto after the common manner; also, the separation of Metals by the force of Fusion, and other necessary things that are done by the power of Fusion or Melting. Most profitable for Chymists, Triers, and Diggers of Minerals.

Of Making the Furnace.

This Furnace may be made greater, or smaller, as you please, according to the matter to be tried: and if the DIAMETER thereof within, be but of one foot, you may set it in a crucible containing two or three pounds; but greater crucibles require a greater furnace. Now this furnace must be quadrangular, and be built of stones, and lute, such which abide the fire, of the height of one or two feet from the bottom to the grate, which must be such as may be cleared from the dross mixed with coals, or such as was the grate of the first furnace, consisting of two strong cross iron bars fastened in the furnace with certain distances for the receiving of 5, 6, or 7 other lesser iron bars which are to be moveable, so that when they are obstructed they may be removed, and cleared from the dross; the lower part of the furnace must have near the bottom a hole (in the forepart) of the height, and breadth of a little span, with an iron or copper door, shutting close: the lower part also must have another hole near the grate on the other side with its register for the government of the fire, and for the attracting of wind. Above the grate, and a hand's breadth from the grate must be another hole for putting in of coals, and crucibles, suitable to the proportion of the furnace, and the height thereof must be of one foot, and the latitude of half a foot, if the Inward Diameter of the furnace be of one foot, whereby the crucibles may be the more conveniently handled, and the coals be cast in with a fire-pan. Let this hole also have a very strong door of stone covered over with lute, either of which may endure the fire, and shut very close, that the fire may thereby (when the crucible is placed in the fire) attract air, but only from the collateral hole under the grate. Let the height of the furnace (being coated above) from the hole appointed for the putting in of coals and crucibles, be of one great span. Let there also be a round hole in the furnace, having the third part of the intrinsical diameter of the furnace, appointed for the flame and smoke, to which if you will use a very violent fire, put to it a strong iron pipe of the height of 5, 6, 8, or 12 feet, for by how much the higher you set your pipe, the stronger fire may you give, and if you will you may erect above the furnace 1, 2, or 3 partitions with their doors serving to divers uses according to the flame that is gathered into them, by reason of divers degrees of fire, which is in them, for the lowest is so hot, that it can easily contain in flux fusible metals, minerals, and salts; and serve for cementation, calcinations, and reverberations; also for burning of crucibles, and other earthen vessels, made of the best earth (of which in the Fifth part) and for vitrification is, and sometimes for trials and burnings, & etc. The second division of heat, which is more remiss, serves for the burnings of minerals and metals as of lead, tin, Iron, and copper, that are necessary for calcinations; also for the necessary calcination of Tartar, and the fixed salt of other vegetables, that is required in chymical operation, as also the calcining of bones, and horns for cupels, and the ashes of wood. The third division or chamber is yet more remiss, and serves for the drying of crucibles, and other vessels that are made of the best earth, and afterwards to be burnt in the first partition. There may also other things be done by the help of these partitions, so that you need not for their sakes kindle a peculiar fire. But if you will give a melting fire the strongest of all, put a long pipe to the lower hole appointed for drawing wind, and having a register; for by how much the fire attracts the air more remotely and another flame is forced to beat upon the metals, so much the greater power of the heat is there in the fusion of them. For which business sake, you need to have as that inferior pipe, so also that superior pipe in the top of the furnace. And if you have a fit chamber, in which another may go up from below by the proper chimney, you may build another furnace in the superior chimney, and perforate the wall with the applying of a register, that the fire may be forced to attract the air from below through the collateral chimney, where you need not that long pipe but only may open a door, or window of the lower chamber, that the air may come into the chimney; and the fire attract the wind out of the collateral chimney, which it does very vehemently, yea and stronger, than if it were helped with bellows, so that even the furnace, unless it were built of very good and fixed earth, would by too great a heat be destroyed; for oftentimes the strongest crucibles melt with too much heat, wherefore a register is made for the governing of the fire.

And by the help of this furnace, with God's blessing, I found out my choicest secrets. For before, and indeed from my youth I underwent the trouble of those vulgar labors performed by bellows, and common vents, not without loss of my health, because of the unavoidable malignant and poisonous fumes, which danger this furnace was without, not only of poisonous and malignant fumes, but also of all excessive heat.

For our furnace sends forth no fume (but above, so drawing), that the door being opened for the putting in of coals, it attracts by the vehemence of the fire, another fume, that is remote by the distance of half an ell. And because the fire does so vehemently attract, it keeps its heat within itself, so that there is no fear of burning; yet you must cover your hand that holds the tongs with a linen glove twice double, and wet in water, and with the other hand a wooden fence that is perspectible to preserve your eyes; otherwise it wants all danger of vapors, or fumes, as has been said, and all excessive heat; the which is a great benefit in Art. I do ingenuously confess, if I had not found this a few years since, I had not without loss left off all Alchemy together with its tedious labors. For I had spent many years of my life in great misery of labors, in superfluous cares, and watching's, as also in stinks, so that going into my laboratory with loathing, I should behold so many materials in so many, and such various pots, boxes, and other vessels, and also as many broken as whole instruments of earth, glass, iron, and copper, and did judge myself so unhappy that I had made myself a slave to this Art, and especially because scarce one of 100, whereof I was one, did get his victuals and cloths thereby. For these reasons, I was determined to bid farewell to Chymistry, and to apply myself to Physick, and Chirurgery, in which I was always happy. But what? While I thought to do as I resolved, and to cast forth of the doors all and each vessel of divers kinds, I found some crucibles broken, and in them many grains of gold and silver, formerly melted in them, which together with others gathered together, I thought to melt; but seeing I could not melt such things being very hard to be melted, without the help of bellows (which I had sold) I began to consider the matter with myself more seriously, and so I found out this furnace, and being invented, I presently built and proved it, which in trying I found so good, that I did again take hope of my labors, and would no more despair.

Seeing therefore an easy, and compendious way of melting metals, I began to work, and to begin a new search, and every day I found more and more in nature, viz. the greatest and most pleasant secrets of nature; wherefore I did without ceasing seek, until God had opened mine eyes to see that which I sought a long time for in vain. Where also I observed, that although I had before had more knowledge of nature, yet without this furnace I could scarce have done anything that had been singular. And so, God willing, by the help of this furnace, I found out more daily, for which blessing I give immortal God immortal thanks, resolving to communicate this new invention candidly, and faithfully for the sake of my neighbor. Judge therefore O Chymist! Whether this, or that which is made by the help of bellows and common vents, be the best? For how long does he that will melt a hard metal in a wind furnace give fire to it before it will flow, and with what loss of time, and coals? He that melts by the help of bellows has need of a companion to blow, with great danger of breaking the crucible with the wind, and of making it fall when the coals are abated, or of impurities falling into the crucible in case the cover thereof should fall off, although there can be no detriment by impurities falling in. If the matter be metallick, but not so if it be a salt or mineral, (without which that cannot be perfected in the fire) not enduring the impurities of the coals, but boiling over because of them. Now our furnace is free from this danger, because the wind comes from beneath and crucibles come always into sight, not being so overwhelmed with coals as in the common way, & etc. For by this means the matter to be melted is flowed, although the crucibles be not covered over with coals, nor with a cover, and although you have not a companion to blow, for you may at pleasure give any degree of fire by the direction of the register. When therefore you make any trial in the fire have this furnace which is recommended to you, which build rightly with its register for the governing of the fire, and for the drawing of wind, and without doubt this labor shall not be in vain.

How Minerals are to be Tried.

The manner of trying minerals has been already made known, wherefore it is not needful here to write many things, because diverse Authors, as GEORGIUS AGRICOLA, LAZARUS ERCKER, and others have sufficiently written thereof, to whose writings I refer you, especially to that most famous LAZARUS ERCKER, which is so much commended, **DE PROBATIONE MINERALIUM**, as well malignant (obstinate) as mild.

But this much know, being that which experience has also taught us, that neither he nor his predecessors had a perfect knowledge of all things, nor would reveal all things they knew. For many excellent things do yet lie hidden, and perhaps shall yet for a while lie hidden by reason of the ingratitude of the world; although the most famous Philosophers do with one consent affirm that imperfect metals, as lead, tin, iron, copper, and Mercury, are intrinsically gold, and silver, although it may seem very improbable to many that are not curious, but contented with the opinions of their parents; supposing those minerals to be barren that leave nothing in the cupel, when they are tried with lead: when as yet that proof by cupels although famous, is not yet that true Philosophical trial of metals, but only vulgar, according to the testimony of Philosophers, as of ISAAC HOLLANDUS[18], and others, especially of PARACELSUS in many places treating of metals, but especially in his book VEXATIONUM ALCHYMISTARUM, containing a true description of the properties, and perfection of metals. Which although not being understood by all, matters not; for a very easy art is not to be communicated to all, according to PARACELSUS. Imperfect metals being freed from their impurities have in them an abundance of gold and silver. But how metals are to be purged, and separated he does not teach, but only commends LEAD to be the Author; which made the Alchemist believe that it was common lead, not knowing that the water thereof (lead) did not only purge other metals, but also lead itself; supposing also that the trial of tin, copper, and iron, made in a cupel with lead to be that true genuine bath thereof; not observing that lead has no affinity with iron, and tin in a

---

[18] R.A.M.S. Library of Alchemy Vol. 26, ***The Mineral Work***. -PNW

stronger fire, but to reject what is black, and unclean, without any perfection. Now this lead can do, if viz. it be mixed with a mineral that has gold or silver in it, and be melted in the fire incorporated with it, it may together with their impurities enter into the Cupel, the good gold and silver being left in the Cupel, which is the proof of minerals that are digged, and used; and it is done upon this account, viz. gold and silver may be naturally purged of their superfluous sulphur so as never to be any more radically united, and mixed with those that be imperfect, as being polluted with abundance of crude, impure sulphur, although they may be melted together in the fire; yet that mixture being retained in the fire, the combustible sulphur of common metals, acts upon its own proper argent vive, and turns it into dross, which being separated from the metals enters into the porous matter of the cupels, that which does not happen in tests, fixed in the fire, which that dross being separated from the metals cannot enter into, being made of an earth that is durable in the fire, the dross remaining in them, which otherwise was wont to enter into those cupels that are made of the ashes of bones, or wood.

    Wherefore by little and little it goes away into the cupel, viz. as much as the fire reduced into a Litharge, or dross, until all the Lead mixed with the Gold and Silver together with other imperfect metals mixed with it go into dross, and hide themselves in the cupel, the pure gold and silver being left in the cupel. For Lead in a plain vessel, feeling the heat from above, but beneath cold, is turned into Litharge, which if it be in an earthen fixed vessel, the Litharge remains, and goes into a yellow transparent glass at last, if it be not mixed with other metals, as iron, copper, tin;

which being mixed therewith, give to the glass a green, red, black, or white color, according to the quantity of the metallick matter: but in a porous cupel made of ashes, the Litharge, or dross finding pores, enters into the cupel by little and little, and successively, until all the Lead be entered in, which could not be if it were not turned into Litharge. This vulgar trying is therefore nothing else but a transmutation of Lead, with the imperfect metals mixed with it, into dross, which entering into the cupel leaves in the cupel pure gold, and silver, that cannot be turned into dross by reason of their purity.

But perhaps this discourse may seem to you unprofitable, and superfluous, because this trial of metals is known all the world over: but for answer, I say that it is not superfluous, because many refiners err, supposing that corporeal Lead together with the imperfect metals that are mixed with it, goes into the cupel, not being yet turned into Litharge, because corporeal Lead is again melted from thence; for whose sake this discourse is not properly ordained, as being those that operate out of use, and custom only without discretion; but rather for their sakes, who do incessantly seek after, and search into the secrets of nature, viz. seeking after that Philosophical trial, which is known to few, by the help whereof more gold and silver is obtained than by the common way, but it is not to be discovered in this place; for all must not have the knowledge thereof. It is sufficient that I have demonstrated the possibility thereof. Yet know this, if you know how to prepare Lead, Tin, Copper, and Iron, and to fit them for a radical union, viz. that aforesaid water of Saturn, so as they may endure the force of the fire together, you may separate and attract gold and silver from the aforesaid imperfect metals, and with gain leaving them in the cupel, or else you shall draw little or nothing from thence (See ***Explicat. Mirac. Mundi.***) And if you do intend to try them with Lead after the vulgar way, and bring them into dross, yet you do nothing, because tin and iron abounding with gold and silver, do not remain with the lead in a strong fire, but are lifted up like a skin or dross, by reason of their superfluous sulphur, swimming like fat upon water, without any separation, unless it be tin or iron, which got gold or silver from the mine in their first fusion.

And by this means it falls out sometimes, that some may make a good proof, but out of ignorance, not knowing a reason of their operation, wherefore they cannot do the same again. For if Chymists, and Refiners did consider the matter more profoundly, enquiring the cause, wherefore lead being tried, deprived of its silver, and melted in a cupel, should yet contain in itself silver, without doubt they would hit upon a good foundation; without which knowledge all their labor in imperfect metals would be in vain, And let this suffice concerning that Philosophical trial, which is known to few; There is no need of speaking anything of that vulgar, being everywhere known, of which LAZARUS ERCKER wrote plainly and fully.

There is also another proof of minerals, which is without Lead, with Venice, or any other good fusible glass, where one or two ounces of the powdered mineral are mixed with half an ounce of the powdered glass, and being mixed and covered in a crucible, are melted, and poured out; by which means the glass attracts, and dissolves that mineral, and is thereby colored, which shows what metal is contained in the mine, after which may be made another trial by Lead, trial being first made by the first proof. And this is the fittest proof for the hardest minerals, which are even invincible, as are the LAPIS HEMITITIS, SMIRIS, granats, talck black and red, and those which abound oftentimes with gold, and silver, which because they cannot be mixed with Lead are not esteemed, but are oftentimes cast away, although they abound with gold and silver, and this because they cannot be tried. Which being tried after the aforesaid manner, and consequently the treasures lying therein being discovered, you may afterward with more confidence handle them, and reduce them to better profit. Now those colors which follow, indicate the tenure of them. Glass resembling the greenness of the Sea signifies mere copper, but the greenness of grass, signifies copper and iron mixed together: glass of a rusty color signifies iron: yellowish glass signifies tin, glass of a yellow golden color, of like a red ruby signifies silver: Blue glass like a sapphire signifies pure gold; a smaragdine signifies gold mixed with silver. An Amethyst color signifies gold, silver, copper, and iron mixed together. Besides these, glass sometimes gets other colors, according to the diversity of the weight of divers metals mixed together; which use will teach with a further practice that is to be made with Saturn.

There is also another precursory trial of minerals, and metals, which is made with Salt-peter, where especially tin, Iron, and copper do largely draw forth their treasures hid in them, which they will not yield being tried by Lead, the which is not a sign of their poverty; but rather of not a true trial made by Lead, which is not the true, and genuine judge, and tryer of metals. For otherwise (if it were) it would draw forth their treasure as well out of a greater quantity of metallick matter, as out of the lesser. Now follows the trial by Nitre: Make a mixture of one part of sulphur, of two parts of pure Tartar, and four parts of purified Nitre, then take an ounce of this mixture, and one dram of the mineral or metal ground small, mix those together, and being put into a crucible, put a red-hot iron or burning coal to them, and that mixture will be inflamed, and yield a most vehement fire, reducing that mineral or metal into dross. And what is not brought into dross must again be mixed with the aforesaid mixture, and be burnt as before, until the whole be consumed by the fire. Then make that dross or salt containing in it the metal that is destroyed, to flow so long in a strong crucible, until it be made glass; which being poured out there are found grains of gold or silver, which came from the mineral or metal that was tried. And this operation (if it be well done) will be a pleasant sight, but without profit, because it cannot be done in a great quantity, and because of the price of the Nitre. Wherefore I set this way of trial only for demonstration sake, that it might appear how almost all tin, iron, and copper, contain in them gold and silver, although they do not draw it forth in the Cupel.

Now do not suppose that this is transmutation, which is only separation; wherefore you must consider yourself how that may be performed otherwise. But take heed that you do not kindle this mixture from beneath, being put upon the coals, but from above, because of the danger of flashing. Also, metals are easily fusible by the following mixture. Take one part of the saw—dust of the wood of the teil—tree being well dried, two parts of sulphur, eight or nine parts of pure Nitre. Make STRATUM SUPER STRATUM in a crucible, and take to 11, or 12 parts of this mixture; one part of the metal subtly ground, and kindle them, and the mine being melted will yield grains of pure Gold and Silver, if the mine were not too impure, the impurity thereof be consumed by that most vehement fire. And if this be not for your profit, yet it is rational, and may be for your instruction.

Of the Melting of Mines and Metals.

The melting of these in a great quantity is not for this place, because they cannot be done by this furnace, but it is treated of plainly enough by others in their writings of minerals.

Of the Separation of Metals.

This is a most ancient and profitable Art, whereby one metal may be separated from another: And it is the most part done always, viz. by AQUA FORTIS, by cement, by flux with sulphur, and lead, and lastly by Antimony; which ways that most witty LAZARUS ERCKER, has clearly, and distinctly described, whose description is not to be found fault with, although some necessary things may be added thereunto, which being but few, I thought it superfluous to add them in this place.

And that separation consists in three chief metals, Gold, Silver, and Copper; he made no mention of other metals, and two of the aforesaid four ways are in use, as very easy, for they are done with AQUA FORTIS and Cement, the two others most commonly neglected, which are done by benefit of melting with Sulphur and Lead; and also by Antimony: that which is admirable, because metals are easier separated by benefit of these two ways, than by AQUA FORTIS and Cements, suspected of waste, whereas not Sulphur and Antimony, but the ignorant worker, not knowing the nature of Sulphur and Antimony, is rather to be blamed, because he knows not how to order them, and withal leaves the nearer way of separation: and I must needs confess it that without this furnace I would not separate with them, because with that common way of furnaces and bellows, the stink of Sulphur and Antimony hurtful to the Liver, Lungs, Brain, and Heart, is received by the Nostrils to the hazard of health: for which cause I do not wonder that those two ways requiring greater diligence than those two former by AQUA FORTIS and Cements are rejected. But this furnace being known, with which without danger one may melt, I doubt not of excelling the two former ways hereafter as more profitable than them. For he who knows Antimony, may not only easily with small cost separate Gold from the addition without any loss of it, and speedily refine it, but also easier separate gilt silver, then by Sulphur, Lead, & etc. in great store without any loss of Gold or Silver.

    And this is the easiest way of the separation of Gold and Silver which is done by the benefit of melting, requiring no more charge than the coals; for there is Antimony which has Gold in it as much as it is worth, which will be the separators gain. I would have you know this, how Antimony may again be separated from Gold and Silver, not by the common way, which is done by bellows, but by the special way of separation wherewith the Antimony is preserved, so that it may be used again for the same purpose; which I will treat of in another place. Besides the four ways spoken of, there is also another way, best of all, by the nitrous spirit of salt, namely after this manner: Rx. the spirit of salt (prepared by our first and second Furnace) actuated with Nitre dissolved in it, to which add grain Gold mixed with Silver and Copper, put it in a glass vial in hot sand to dissolve, and the Gold together with the Copper will be dissolved in it, and the Silver left in the bottom of the vial. Decant off the solution, to which add something, precipitating Gold, and make them boil together, and the pure fine gold will be separated and precipitated like the finest meal, serving Writers and Painters; the Copper being left in the water; which you may if you will precipitate from the water, but it is better to take away the water, which will serve again for the same use. If the precipitated Gold be washed and dried it gives in the melting (by which nothing is lost) the best and purest Gold. For finer gold, can neither be made by AQUA PORTIS nor by Antimony.

    Therefore, this is the best way of all, not only for the small cost, but also for the easiness yielding the best Gold of all others.

    Then take the calcined Silver left in the gourd, sweeten and dry it, which done make a little salt of Tartar to melt in a crucible, to which by course put a little of the refined silver with a spoon, and it will presently be made a body without any loss. You may also boil that Calx yet moist newly taken out of the gourd with a Lixivium of Salt of Tartar, even to the evaporation of all moisture: and melt the dry remnant, where also nothing is lost. Without this medium the calx of Silver (drawn from AQUA REGIA) is not fusible of itself, turning into a brittle matter, like horn that is white, or of a middle color between white and yellow, called therefore of Chymists, the HORN OF THE MOON; in reducing which many have tried much, which reduction we have already taught. For want of spirit of salt take AQUA REGIA made of AQUA FORTIS and salt Armoniack, which does the same, but with greater charges. This also is to be preferred before other ways, which makes to the separation of any Gold of any degree, if so be it exceed Silver in weight; which is necessarily required in the solution made with AQUA FORTIS.

But that you may see the prerogative of this separation, mark a little, when you separate by the QUARTO and by AQUA FORTIS you must put just two or three parts of refined Silver to one of course Gold, where first the cost and labor of refining the Silver to be melted and grained with Gold are required: then a good quantity of AQUA FORTIS to dissolve, precipitate, edulcorate, dry and melt a great deal of silver. Consider then I pray, the labors and charges of my separation and the vulgar. When you separate with Cements there is need of boxes, and continual fire of one degree, which labor is tedious for time's sake, and costly for coals, which labor you must twice or thrice take in regard of the mixt dross. Now again consider the labor and charges of both separations. When you separate by Sulphur and Antimony, which is the best way, without great charges, if you know to separate Gold from Antimony without blowing, but this is tedious because thrice greater labor, then our way, tedious indeed because of the difficulty of a perfect separation of Gold and Silver from the Antimonial dross. Think therefore what way of separation you will use to refine Gold speedily, surely you will choose mine.

This way of separation has also this prerogative, that it has no need of refined silver which is done by the benefit of burning, but only its granulation, solution or separation using AQUA FORTIS, where though copper mixt with silver makes waste, yet by the help of this salt it is soon precipitated. By this means gilt silver is soon separated, the gold being dissolved by the nitrous spirit, and precipitated with the aforesaid matter precipitating. As for the separation of gilt silver which is to be done by the help of fusion, and none is easier done than with Sulphur and Antimony, where when the necessary manual (ingredients) are known; a great deal is separated in a short time, but if you know not how to handle Antimony and Sulphur (for which our Furnace very well befits) leave there, and use the common way, therefore lay not your fault afterward on me, writing for your good.

Of Separating the Courser Metals.

The manner of separating Tin from Lead, and Copper from Iron, without loss of both metals, by preserving both, has hitherto been unknown, which seems impossible to me because of the combustibility of both metals; and superfluous for the small profit, and saving charges. But how Gold and Silver may be separated from Tin with which commonly this abounds, without any waste, has been long since sought to no purpose: but a possibility will appear to a serious considerer; and though I never tried in great quantity, being content with a precipitation made with a little; I am yet persuaded this business will succeed in a great quantity and with much profit; namely by the help of a Furnace made on purpose where gold and silver precipitated with lead and HALB KOPF[19] by extreme heat of fire; that tin is extracted to the remanence of the tenth part, which remainder you must peculiarly take and keep. Which done you must precipitate new tin in the foresaid Furnace, and so extract to the remainder of the REGULUS, which being extracted from, is to be added to the first and reserved; which labor is to be reiterated, till you have a sufficient quantity of REGULUS filling the Furnace; which again you must precipitate; for by this means gold and silver are brought together, so that they may easily afterward be separated from the superfluous tin. By this means I count the separation profitable, where but little substance is lost, which is turned into ashes and smoke. Nor does adding lead and HALB KOPF hinder, because sometimes lead is mixt with tin, and the HALB KOPF is separated again. It is good therefore to separate pots and old dishes, because of the mixture of lead, and to precipitate the gold and silver from them, by the adjection of HALB KOPF only, where the residue is no way altered by the HALB KOPF, therefore you may sell it, or refine it again: which in my judgement will be to great advantage.

What is to be held concerning the Perfection of Metals.

This knot is scarce soluble, for so many and divers opinions of so many ages, so that most men slighting the testimonies of true Philosophers, will not believe the truth, especially because scarce one of a hundred can be found who is not impoverished with this art: the incredulous therefore is not to be blamed for his doubting, no signs of truth appearing, yet experience testifies a possibility by

---

[19] "Half head" in German. -PNW

art and nature, though examples are rare, I pray with how great absurdity should one deny Heaven and Hell never seen? But you say we must believe this as revealed by God, his Prophets and Apostles; but so is not this, but the Philosophick tradition of Heathens. I answer, though most Philosophers were heathen (yet some have been Christians) yet their works are not to be despised, because not handling our salvation: to whom if CHRIST had Preached, surely, they had believed him. For it appears by their books that they were pious and honest men; who though not Professors of CHRIST, yet they did His Will indeed which we, though not in words, in action deny; who if they had been wicked, why took they so much pains in making books for the good and profit of their Neighbor about Virtue and Piety? Why spent they not rather their life time in leisure and pleasure, as is the custom nowadays with them who are appointed to instruct us? Why should they gull posterity with trifles and lies, expecting from thence no profit? For most of them were not poor, but very rich Kings and Princes. Besides these, there have been many Christians seriously confirming the truth of the Art: Men indeed of special note, namely, Bishops, Doctors, & etc. Such were THOMAS AQUINAS, ALBERTUS MAGNUS, LULLIUS, ARNOLDUS, ROGER BACON, BASIL, & etc. Why should very pious men deceive posterity with their Works, and lead them into Errors? Although there should not remain the Works of Famous Worthies, yet there would be a plain confirming the truth of this Art. For I am persuaded there are some to be found having this knowledge, and privately possessing it. For who is so mad to reveal himself to the world, to receive nought but envy for his reward? Let no man therefore doubt of this secret Art's truth. But say you: Why stand you so much for the Art? Did you ever see or perform anything in it? I reply, though I never made projections to perfect metals, nor saw transmutations; yet I am sure of this, I have often from metals with metals, leaving no gold and silver in the cupel, extracted gold and silver by the help of fire.

But I will not have you think that one imperfect metal will perfect another, or turn it into gold or silver, impure and drossy without, in comparison of gold and silver; for how can such metals perfect another imperfect? Which thus understand. For as in the vegetable Kingdom, water cleanses water, or juice with seething as is wont to be done in purifying honey and sugar, or any other vegetable juice, with common water, and white of eggs: so also you must understand of mineral juices or metal, of which if we know the water and white, surely we might refine the impurity, in which gold and silver lie hid, as in black shales, and powerfully extract gold and silver, which is not a transmutation of metals, but an eduction of gold and silver from the dung-hill. Do you ask how Gold and Silver can be educed from copper, iron, tin, and lead, to wit, by the help of lotion, out of which none is drawn with that best proof (as it is thought) of Cupels? To which we answered before of the proof of Cupels not to be sufficient for all the several metals. I need therefore say no more, but I refer the studious Reader to PARACELSUS his Book, the **VEXATION OF CHEMISTS**, where you shall find another lotion and purification of metals, which heretofore was unknown to Miners and Dealers in Minerals. As for example: A Miner finding the ore of copper, uses his skill delivered by the ancients to his utmost endeavor, whereby he may cleanse it and reduce it to metal: where first he breaks it into pieces, and boils it, for to take away the superfluous Sulphur, then by virtue of melting, he brings it into a stone (so called) which afterward again he commits to fire, and frees it by the addition of lead, of its gold and silver: which done, he blacks and reddens it, turning it into copper, which is his last labor, whereby the copper is made malleable and vendible: which done, the Chymist coming, tries another separation, by whose help gold and silver is extracted, as yet tried of very few, of which mention is here made. PARACELSUS also says in the same place, that God has given some an easier way of separating gold and silver from courser metals, and indeed without refining the ore, which is a special and curious Art, which he teaches not in plain terms, but only says it is sufficiently taught in seven rules of that book, where he treats of the nature and propriety of metals; in which you may seek it. And this purification of courser metals I count most easy, which I have often tried in small quantities: and I doubt not but God has shown other Artists also other purifications by which imperfect metals are perfected; for example, if one would purge the fruit of the earth by distillation, so that the dregs and impurities being taken away, it would grow up with a new clear clarified Body: as if one distil black and impure Amber by a retort, the separation would be made by Fire, of the water savoring of an EMPYREUM, of the oil and volatile salt, and the CAPUT MORTUUM be left in the bottom of the retort; by which means, in a very short time without great labor, is made a great alteration and emendation of Amber, though the oil be black, impure, and stinking: but if it be again distilled by a retort with some mundifying water, as with the spirit of salt (namely through a fresh clean retort) there will be made a new separation by that spirit of salt, and a far clearer oil will be extracted; the dregs with the stink left in the bottom of the retort, which afterward may be twice or thrice rectified again with fresh spirit of salt, until it get the clearness of water, and sweetness of scent resembling Amber and musk.

And this transmutation makes of a hard thing, a soft, unlike the former in shape, which though never so soft and liquid, oily, may again be coagulated, so that it becomes as it was at first, after this manner following. Take the said oil very well clarified, add to it fresh spirit of salt, set it in digestion, and the oil will attract from the spirit of salt, salt enough for its own recoagulation, and again it acquires the hardness of Amber, of an excellent clear and admirable color; of which half an ounce is worth more than some pounds of black Amber; of which scarce the eight or tenth part remains in purifying, all the foul superfluities cast away.

By this means I think one may cleanse and mend black metals, if so be the manner of their cleansing were known by distillation, sublimation and recoagulation. But you say that metals cannot like vegetables be purified by force of distillation, to which I present our first furnace not given to peasants, but Chymists, purifying metals; so also, the possibility of their perfection is shown by help of fermentation. For as fresh leaven can ferment the vegetable juices, which are perfected by fermentation, the dregs being cast away as one may see in wine, ale, and other liquors, whose lasting and perfection proceeds from no other thing but fermentation purifying the vegetable juices, without which they could not otherwise withstand the Elements, subject to corruption in a very short time, which fermented last some years: so also if we knew the proper ferment of metals, surely we might refine and perfect them, so that they not being any more subject to rust, would be able to prevail against fire and water, and be nourished and fed by them. For so the world heretofore perished with water, and shall at last perish with fire, and our bodies must rot and be purified by fire before we come to the sight of God. And thus, far of the fermentation of metals, wherewith they are amended and perfected. Metals also are purified and amended like milk set on the fire; whose cream the better part (the substance of butter) in the top is separated from the whey and cheese, and the hotter the place is, the sooner the separation is made even, so it is with the separation of metals; where metals put into a fitted hot place by themselves without any addition of another thing (the metals being before reduced to a milky substance of curd) are separated in time, by parting the nobler parts from the ignoble, opening a great treasure: and as in winter time milk is hardly separated with a weak heat; just so metals if not helped with Fire, as one may see in iron, which in a long time under the earth is turned into gold without Art. For often iron ore is found with golden veins very goodly to behold, severed from the course, earthy and crude sulphur, by force of the central heat. And commonly in such ore no vitriol is found, being separated and bettered by its contrary. But a long time is required for that subterraneous separation, which Art very speedily performs; as is wont to be done with milk in winter when we presently make butter of it, when we put it to the fire to part the cream speedily; which separation is helped by the precipitation made with acid things, mortifying the urinous salt of the milk, by which means all principles are separated by themselves, as butter, cheese, whey: so in a quarter of an hour separation is made by boiling, which else without acid things could not be done in some weeks. If then it be possible in vegetables and animals, why not in minerals? For what but gold and silver is found in lead, iron, tin, and copper, though it does not appear? Why is all goodness denied to the courser metals granted to vegetables and animals not equal to them for lasting? Whence is the natural perfection of lead, tin, iron, and copper to be proved? Nature ever seeks the perfection of her fruits: but course metals are imperfect: Why then is not nature helped with Art in perfecting them? But the bond of metallick parts is worth observation, which being broken, the parts are separated. Urinous salt (as I may say) is the bond of the parts making milk, as of butter, whey, and cheese, which is to be mortified by its contrary acid for separation. But in iron the parts are bound with a vitriolate salt, as with a bond, which is to be mortified with its contrary, urinous or nitrous salt for separation. Be therefore who knows to take away the superfluous salt of iron, either by moist or dry means, doubtless shall have iron not soon subject to rust.

    Fire also has incredible force of itself in changing metals. Is not steel made of iron by force of Fire, and iron steel by different proceeding? Experience daily teaches us also divers kinds of changes and refinings by Fire; why is it not possible in metals by an expert Chymist having skill in them? Who would believe that a live bird lurks in an egg, and an herb having leaves, flowers, and odor, in the seed? Why may not then abortive metals, getting not yet perfection, be perfected by Art, with help of Fire? Is not an unripe apple or pear ripened by the heat of the Sun? Which some curious and industrious men observing, have imitated nature in their works; and have found some metals not destroyed with the heat of Fire, but enriched with a secret gainful heat: so, that melted (digestion being made) they have yielded double weight of gold and silver. Yea I myself have seen the common ore of lead digested after the aforesaid manner, which was not only enriched with silver thereby, but also partook of gold which it wanted before in ordinary trial. Besides one might work this in great quantity, as with a hundred pounds; which work of minerals will without doubt bring great profit to the skillful triers of lead. But know this, that not every trial of lead will be furnished with gold, but the ore to be ever enriched with silver, experience being witness.

   Many such things are found in Nature, incredible to the ignorant, and those that are unexercised. But if we mortals were more diligent in reading the book written with the hand of God in the pages of the four Elements, surely, we should find more secrets and wonders in them, but skill and wealth is got with sweat of face and not by sloth; therefore LABOUR and PRAY. Metals are also meliorated by the help of gradation like unto germination.

   For it is well known, that the shoot or grass of some fruitful garden tree implanted in a wood, makes that tree afterwards to bear not wild fruits, but very good and sweet like them of the implanted shoot, as one may see in iron dissolved in an acid spirit, fermented with Venus and turned into Copper:

by which means doubtless copper is turned into silver, and silver into gold, if the true manner of fermentation were known.

Now this transmutation is like digestion, making beef or horse flesh of grass in the stomach of oxen and horse, and man's flesh or beef, in the stomach of man.

The better parts also are separated from the worse parts by the attractive strength of the like, as is to be seen in a metal abounding with sulphur, to which if iron be added in fusion, the sulphur deserts its native metal, (by which means it is more purified) and joins itself to the iron, with which it has more affinity and familiarity, than with its own metal; for example, if iron be added to lead ore full of sulphur in the melting, this melted metal is made malleable, which else would be black and brittle. And if something else to be put to the melted malleable metal were known to us, to take away in the melting, the redundant, crude, combustible sulphur, question less it would yet be made purer; which thing being unknown, metals remain in their impurity. And indeed, God has done well in this as in all other his works, that He has concealed His knowledge from us: for if it were known to the covetous, they would buy up all lead, tin, copper, and iron, to turn into gold, so that rural and poor Laborers could hardly buy metallick instruments for their use, for the scarcity; but God will not have all metals turned into Gold.

A Similitude of taking away the superfluous sulphur of some metals in fusion, being given to keep the purer parts; so likewise, is there another manner of separating, the purer parts from the impure, namely, by the attractive power of the like, where the purer parts are drawn together by their like, the impure and heterogeneous part is rejected: and that may be shown as well by the moist as dry way: an example of the moist way follows.

If quick Mercury be added to impure gold or silver dissolved in its proper MENSTRUUM, the mercury draws to itself the invisible gold and silver from the MENSTRUUM and mixt Impurity and associates what is purest to itself, which separation swiftly succeeds. Mercury performs the same likewise in the dry way: namely, when some earth having some gold and silver, is moistened with acid water, and they are so long bruised together, till the Mercury draws the better part; which done you must wash the dead earth left, with common water, and separate the mercury being dried from the attracted gold and silver, by trajecting them through a skin, but the Mercury draws but one metal from the earth, and indeed the best at one time; which being separated, it draws another metal; for example, if in some one earth, gold, silver, copper and iron lie hid, the first time the mercury draws the gold, the second the silver, but copper and iron hardly by reason of their dross, but tin and lead easily, but easiest of all gold by reason of its purity like to mercury.

Another Demonstration by the Dry Way.

Put under a tile a cupel with lead, to which add a grain of very pure gold, most exactly weighed (for memories sake) make the gold in the cupel to fulminate, and the lead will enter the cupel, the gold being left pale in the cupel: of which pale color there is no other cause than the mixture of silver, drawn from the lead by the gold. But you will say, that you know this, that gold fulminated with lead, is made paler and weightier, by reason of the silver in the lead, left with the gold in the trial, augmenting the weight, and thence making it pale: to which I reply, though lead leaves some silver in trying in the cupel, mixt with the gold added to it, augmenting its weight, and changing the color; yet it is proved by the weight, that lead leaves more being mixt with gold in the cupel, than when tried without gold. Hence it is proved, that gold in the fire draws its like from other metals, augmenting its weight: and this also gold does in the moist way: for if it be dissolved in its own MENSTRUUM, together with copper, and put in digestion, and then separated, it attracts gold from the copper; which labor, though not done with profit, yet witnesses a possibility. But if the MENSTRUUM of gold augmenting the attracting power of gold or multiplying the same were known, but diminishing the retentive power of copper, doubtless some gain were to be expected; and indeed more, if gold and copper, together be melted in fire with the dry mineral MENSTRUUM, by which means the weight of gold would be increased according to PARACELSUS saying Metals mixt together in a strong fire, continued a pretty while, the imperfection vanishes and leaves perfection in its place.

    Which surely, if done well, is a work not wanting gain. For I freely confess that I would sometime incorporate silver with iron, when as gold from iron gave me a good increase of pure gold, instead of fixed silver sought after. And by this means often some not thought on thing happens to Artists, as to myself with fixed silver, not rightly considering the business. Therefore, meddling with metals, be sure when you find some increase, to weigh well what it was at first. For many think long trying silver with iron, by the Blood-stone, Load-stone, Emerald, LAPIS CALAMINARIS, Red-talck, Granats, Antimony, Arsenick, Sulphur, Flints, & etc. having mature and immature, volatile and fixed gold in them, finding in the trying good gold; that this gold is made of the silver by the help and use of the foresaid minerals, which is false. For the silver drew that gold out of those minerals, in which before it lurked volatile. Yet I deny not the possibility of changing silver, as being inwardly very like gold, but not by help of cementation with the said minerals, because that gold proceeds not from the silver, but those minerals, attracted by the silver. This labor is compared to seed cast into good ground, where dying, by its own power it draws its like to itself, whence it is multiplied a hundred-fold.

And it behooves in this work now and then to wet the metallick earth, with proper metallick waters, being dried up with heat (which operation is called of the Philosophers inceration) else the earth will be barren, and it behooves that this water be near in kind to the earth, so that when they are united they yield a certain fatness. For as it appears from sandy dry earth, moistened with rain water, not bringing forth fruit agreeable to its seed, for the small heat also of the Sun consuming the moisture, and burning the seed in the earth, which mixt with cow's dung or other, keeps the water so as that it cannot be so soon consumed. By the same reason, it is necessary that your earth and water be mixt, lest your seed be burnt up. Which work if well handled, it will not be in vain, requiring the exceeding diligence of nourishing the earth with warmth and moisture, when the earth is drowned with too much moisture, or has too little, it cannot increase, and this is one of the best labors with which I draw forth good gold and silver of baser metals, requiring the best vessels, retaining the seed together with its earth, and water in its proper heat. I doubt not but this work also in greater quantity may be performed, firmly believing that the courser metals, especially lead, the fittest of all not only to be perfected into gold and silver, but also into good medicine: which without question is a Philosophick labor granted from God, as a great comfort to the Chymist, but warily to be used. For that all and singular God's gifts he will not have common: as indeed I have found, when I had invented a very excellent work, that I showing it to a friend, neither could I afterwards teach it to him, nor do it again for myself. Therefore indeed justly men are doubtful in writing such matters: for many seek with idleness to get the inventions of others, performed with great costs and labor. Therefore, it is safer to be silent and give leave to seek, than to publish secrets, that they may undergo the pains and charges to be born in inventing high matters; nor any more hereafter may the ungrateful so impudently gape after others Labors. Therefore, I would entreat all men both of high and low degree, that they would not molest and tire me hereafter with their Petitions and Epistles, and that they would not turn my good will of benefiting others to the ruin of myself, but be contented with my writings published for the profit of my neighbor. Nor do you think that I possess and promise golden mountains. For what I have written, I have writ to discover nature, in these discourses of the perfection of course metals in small quantity: For I have never made trial in a great quantity, trying truth and possibility in a lesser only, in small crucibles: therefore those things which I have writ are written to that end that the possibility of the Art, may appear, of perfect metals to be wrought out of imperfect, therefore he who has occasion may make trial in a greater quantity: but as for my part wanting opportunity, I expect God's blessing, whereby upon occasion I may make trial in a greater quantity, and so receive the fruit of my labor and great charges.

Also, metallick bodies are transmuted by another means, namely by the benefit of a tinging metallick spirit, as one may see in AURUM FULMINANS, sometimes kindled upon a smooth clean metallick plate, fixing a very deep golden tincture upon the plate, so that it may bear the Touch-stone. The same also happens in the moist way, where plated metals put into a gradatory spirit made of Nitre, and certain minerals, being pierced by the spirit, obtain another kind agreeing to the spirit. But if one doubts of the metallick gradation, made with AURUM FULMINANS; he may try the certainty from the often firing of fresh AURUM FULMINANS, upon the same plate; for he shall see that it is not the color of the metal, and outwardly gilded, but deeply tinged. Likewise one may try the certainty by a humid spirit, if the transformed metals are tried, whence the mutual action and passion of subtilized spirits plainly appears, for the power of spirits is very great, and incredible to one not exercised; and this gradation of inferior metals, Philosophers both ancient and modern, do not only confirm, but also diggers of minerals taught by experience, that mineral vapors by penetration change courser into purer metals, LAZERUS ERCKER being witness, that iron is changed into good natural copper in green salt waters, & that he saw a pit, in which iron nails and other things cast in, by the penetration of a cupreous spirit were turned into good copper. I do not deny that metallick dissolutions of some metals do stick precipitated to the plates, and to make them of a golden, silver, or cupreous color; for it is well known, that iron cast into a vitriol water not to be turned into copper, but to draw copper out of the water, of which thing we treat not here, confirming the possibility of metallick transmutations by a tinging and piercing spirit; therefore I again maintain that great power is in metallick spirits; look only upon course and opaque earth, and besides that clear and limpid water with which the clearer and more powerful air proceeding from the water comes from the earth. Are not whole Countries drowned with water, sometimes Towns and Cities taken away? Cannot the air destroy the strongest Houses; especially shut up in the earth, shake the Land for some miles, and afterward demolish whole Cities and Mountains with the death of Men? All which things are done naturally. Wind artificially raised by Nitre threatens a far greater danger, which no man can deny. Although that corporeal Elements exercise so great power, yet they cannot pierce metals without hurt, nor stones and glass, and things soon penetrated by fire. Therefore, not by an occult but a manifest power of Sun and Fire, which it has over metals, stones and glass, which are easily pierced by them without any impediment: and why should not metals compact of a certain metallick subtle and piercing spirit be penetrated by help of fire, and changed into another species? As is already spoken of AURUM FULMINANS and AQUA GRADATORIA. Therefore, there is no doubt of the possibility of the metallick tingent spirit changing courser metals into finer, both by the dry and moist way. For Metals may be purified the same way as Tartar and Vitriol, and other salts, namely by the benefit of much water. For it is manifest that vitriol is purged with iron and copper mixt with it, namely dissolved and coagulated in much water, so that it waxes as white as allome; which purification is but a separation of the metal from the salt, made by the benefit of much water debilitating the salt, so that it cannot longer retain the mixt metal, which is precipitated like some slime, not unprofitable, because the chiefest part of the vitriol, from which is the greenness, viz. Copper, Iron, and Sulphur. And as by help of separation metals are drawn from vitriol, more perfect than salts; so also, it is with metals when the more perfect and better part is separated by help of precipitation: as for Tartar, it is nourished by the addition of water, but its better part is not precipitated as in vitriol, but the courser part which is its blackness and feculence. As for example: Common Tartar by the often solution (made with a sufficient quantity of water) and coagulation is made very pure and white, because in every solution made with fresh clear water, it always becomes purer; and not only by this means white Tartar, but also red and feculent, is reduced into transparent crystals, and indeed very speedily by virtue of a certain precipitation; whose limosity is the cause of the obscurity of the crystalline salt of tartar, and is nothing else but an unsavory thing, dead and useless, mixt with the tartar in its coagulation in Hogs-heads of wine, and separated again by power of solution.

    And these examples of the two salts of Vitriol and Tartar, are not in vain set down, because they show the difference in precipitation: For in some Metals, by force of precipitation, the courser part is separated; but in other, the better and choicer, according to the prevalence of this or that part.

In Vitriol, the better part (Copper and Iron) is the least, which is precipitated and separated from the courser and greater part, viz. Salt. But, in Tartar, the courser and less part is precipitated and separated from the greater and better part clarified: The like is in Metals. Therefore, let everyone be wary in separating: and consider before, whether the better or courser part of the Metal is to be precipitated; without which Knowledge, no Man can meddle with this business. Let also the Workman beware, who expects any profit from his labor, of Corrosive Waters; as AQUA FORTIS, AQUA REGIA, Spirit of Salt, Vitriol, Allome, Vinegar, & etc. in the solution from which no good proceeds, as utterly destroying and corrupting all and each of them: proving the same in these words, FROM METALS, BY METALS, AND WITH METALS, METALS ARE MADE PERFECT. Metals are also purified; maturated and separated from their Vices, by Nitre burning up the superfluous Sulphur.

And all the aforesaid perfections of metals are but particular. For every particular medicine, as well humane, as metallick, purges, separates and perfects or amends by the taking away the superfluity. For a universal medicine works its perfections and emendations, by strengthening and multiplying the radical moisture as well of animals as metals, expelling its enemy by its own natural virtue. But you say, excellent examples indeed are delivered by me, but not the manner of doing them. R. I have delivered more than you think, although you don't perceive it: for I am sure after my death that my books will be held in a greater esteem, from which it will appear that I have not sought vain glory, but the profit of my neighbor to the utmost of my power. But do not, seeing my freeness of writing, think that you may wrest many things from me. For assure yourself, that although I have written many things for the public good, yet I intend not by this means to trouble myself. For I cannot satisfy the desires of all men, nor answer their Epistles, nor enrich all men, who neither am rich myself, nor have sought riches. For although I have gotten the knowledge of these things by God's blessing, and have tried the truth of it in small quantity, yet have I never made experience in great store for wealth's sake, being contented with God's blessing.

And let this suffice concerning the several purifications of metals according to my experience; as for that universal medicine so famous, I cannot judge of it, being a thing unknown to me; but the possibility thereof I am forced to affirm, being moved with the several transmutations of metals; which being unknown, it behooves us to be contented with that favor which God has bestowed on us. For oftentimes question-less it is better to know little, for Eternal Salvations sake; for most commonly wealth and learning puff up. And pride brings to the Devil the Author of it, from whence God of his mercy preserve us.

Of the Philosophers Stone.

I have undergone much charge and labor for many years, to extract the tincture or anima of gold, for a medicine to be made therewith, which at length I have obtained, where I have observed the remainder of the gold, the soul or better part being extracted to be no more gold, nor longer to endure fire. Whence I conjectured, that such an extraction being fixed again, can perfect courser metals and turn them into gold. But I could not hitherto try the truth of my conceived opinion living at this time in a foreign place; therefore, against my will, although greedy of novelty, I have been forced to abstain from the work. In the meantime, considering the opinions of the Philosophers concerning their gold, not the vulgar, asserting the universal medicine to be prepared therewith. I have again affused a certain Philosophical Vinegar to Copper for to extract the tincture, where almost all the Copper like whitish earth is separated from the tincture in digestion, which earth by no Art I could again reduce into a metallick body.

    Which experiment again confirmed me of a possibility of this Medicine. Which labor though I never followed, yet I doubt not but a humane medicine, though not also a metallick is attainable thence by a diligent workman. The soul therefore with all the metallick attributes, consisting in so small a quantity, which is scarce the hundredth part of the weight, which being extracted and separated, the remaining body is no more a metal, but a useless and dead earth; but it is not to be doubted but being fixed again, it may reassume and perfect another metallick body. Therefore, I am confidently persuaded by the aforesaid Reasons, that such a medicine is to be made of mineral and metallick things, viz. in the flowing, changing baser metals into better. But do not think that I writing these things make gold or copper the matter of this medicine, which I do not hold, well knowing that there are other subjects easily to be handled, abounding with tinctures.

So, you have heard now my opinion of the Universal Medicine, which my experience in Gold, Copper, and other Minerals and Metals has caused: which I will not preach for GOSPEL, because it is human to err. Therefore, no certainty is to be had, before its final and complete perfection, and indeed once or twice tried for certainties sake. For an excellent way once found out, cannot always be often repeated, which happens doubtless as well to others as to me. Therefore, we must not triumph before the Victory; for unthought on impediments may frustrate Hope: but God is rather to be implored in our labors, that he would be pleased to bless our endeavors, that we may use well his gifts in this life as good stewards, and afterward bestow the free reward of our labors, watchings, and cares on us sinners, namely, everlasting Rest and Salvation out of his mere Mercy.

Whether Minerals, as Antimony, Arsenick, Orpiment, Cobalt, Zinck, Sulphur, & etc. may be transmuted into metals, and into what?

It is long since debated among Chymists, whether the aforesaid Minerals proceed from the same principles with Metals, and whether to be counted Metals; in which Controversy they have not agreed to this day, when as one approves that which another denies, so that a student of Chymistry knows not to what side he had best assent.

But this knowledge not a little helping, concerning the purifying of metals, I would put my opinion also grounded upon experience, for the satisfying the doubtful, the simplicity of them is strange who hold not one and the same beginning to be of minerals and metals, saying, if metals might be made by nature, of minerals surely it had long since been done; but it was never, experience witnessing; for remaining minerals, they are never transplanted into metals. I answer, metals grow one way, also vegetables another, soon budding, and again soon dying; but it is not so with metals; for all lasting things have a long time of digestion, according to the saying, THAT WHICH IS SOON MADE, DOES SOON FADE; this is to be understood not only of vegetables and minerals, but also of animals, as appears from the budding of some vegetables, coming in six months space to their perfection, and then again perishing: when as things requiring longer time of digestion and perfection are much more lasting. A Mushroom in the space of one or two nights grows out of a rotten wood, again soon vanishing: not so the Oak. Oxen, and Horses in the space of two or three years come to perfection, scarce living the twentieth, or twenty-fourth year: but a Man requiring twenty-four years to his perfection, lives sixty, eighty, or a hundred years. So also, we must conceive of lasting metals requiring many ages, and a very long time of digestion and perfection. Metals therefore requiring a very long time of digestion to their perfection, it is granted to no man ever to see the beginning, and the end of them; the transplantation of minerals into metals by nature cannot be denied; especially, because that in the ores of metals, especially of course ones, minerals are also found; wherefore diggers of minerals, when they find them, conceive good hopes of finding metals, of which they are termed the COVERLIDS, for seldom metals are found without minerals, or minerals without metals; nor also are ever minerals found wanting gold or silver; therefore minerals are properly termed the EMBRYO of Metals; because by art and fire a good part of gold and silver is drawn out of them by fusion; which if they do not proceed from the metallick roots, whence proceeds that gold and silver? For an Ox is not born of an infant, nor a man of a Calf, for always like is produced of its like.

Therefore, minerals are counted but unripe fruits in respect of metals, not yet obtaining their ripeness and perfection, nor separated from the superfluous earth; for how should a bird be hatched of an egg by a heat, not predestined for the generation of a bird? For so we must understand of minerals, which if they be deprived of their metallick nature, how should by fire metals be produced from thence? But you say that you never saw the production of perfect metals out of courser; therefore that it is neither likely, nor credible to you, to whom many things as yet lie hid, as from most men, perversely and foolishly denying things unknown; for daily experience witnesses, that the viler minerals and metals by taking away the superfluous sulphur (however it be done) obtain a greater degree of perfection, therefore should not your heart believe, and your tongue speak what you see with your eyes? For experience shows that good gold and silver might be drawn out by art almost out of all course minerals and metals, yet more out of some then of others, and speedier; for there is not that dark night, that is altogether deprived of light, which may not be manifested by a hollow glass; nor is there an element (though ever so pure) not mixt with other elements, nor any malignity deprived of all good, or on the contrary. And as it is possible to gather the hidden beams of the Sun in the Aire, so also hidden perfect metals dispersed in the imperfect metals, and minerals by fire, and an expert Artist: if once they are placed in fire with their proper solvents, where the homogeneous parts are gathered, and the heterogeneous separated; so that there is no need to go into the INDIES to seek gold and silver in those new Islands, which is possible to find plentifully here in GERMANY, if so be the merciful God would please to turn away those present cruel Plagues, and bring them out of old metals, viz. Lead, Tin, Iron, and Copper, there left by the Dealers in minerals; indeed without the culture of minerals. Let no man therefore judge himself to be poor, because he is only poor and in want (although otherwise very rich and abounding in wealth, which yet in a moment be is forced to forsake) that being ungrateful, neither knows nor acknowledges God in his works.

What, I pray, is in less esteem in the world, than old Iron and Lead, which are acceptable to the wise to use in the Lotion of Copper and Tin with the mineral White? But how they are to be washed, is a difficulty to the unexercised in the fire, and shall be delivered by similitudes: You see Antimony fresh digged out of the earth, very black and impure; which by fusion separated from its superfluity (which, though nature gave to it not in vain, but as a help to its purification, according to that: GOD AND NATURE DO NOTHING IN VAIN) is made more pure, and endowed with a body nearer to metals than its mineral, which if afterwards melted with the salt of Tartar, the crude and combustible sulphur is mortified thereby, and is turned into dross, and separated from the pure mercurial part, so that hereby is made a new and fresh separation of the parts, of which one portion being white and brittle, sinks to the bottom, the other lighter, to wit, the combustible sulphur is on the top with the salt of Tartar; which poured out into a Cone, when they are cold, may be separated with the hammer; the inferior part of which is called by the Chymists REGULUS, which is purer than Antimony cast the first time out of its mineral; and this is the usual purging of Antimony used by Chymists; to which (REGULUS) if afterward anything should be added, for a third purification, without doubt it would not only be made purer but more fixed and malleable. For if white REGULUS be can be prepared out of black Antimony, why not as well malleable metal out of the REGULUS.

Another Way of Separating the Superfluous Antimonial Sulphur.

Rx. Antimony powdered one part, Salt-peter half as much, mingle them, and kindle the mixture with a live coal, and let that Antimonial sulphur with the nitre be burnt up, the darkish mass being left, to wit, of a brown color; which melted for the space of an hour in a strong fire yields an Antimony like to that which is made with salt of Tartar, but somewhat less in quantity: in like manner the parts of Antimony are separated, viz. if Antimony, Nitre, and crude Tartar be mingled in an equal weight, and being mixt are kindled and melted. There is also another separation of the Antimonial parts; when of small bits of iron one part is put into a strong crucible, in a wind Furnace, to which being red hot, cast two parts of ground Antimony, for fusion, and the superfluous combustible sulphur will forsake the Antimony, and join to the iron, a metal more amicable to it; mixt with which, it forsakes its own proper pure Mercury, and sulphur or REGULUS, which is almost the half part of the Antimony.

And these four ways, by which the superfluous combustible sulphur of Antimony is separated are most common, not set down as secrets, but for demonstration sake, that it may appear how sulphureous minerals are to be perfected and purified, which are little amended; yet shewing a better way not only for Antimony, but also for Arsenick and Orpin[20], although these two cannot be so done with Iron, Nitre and Tartar because of their volatility; but with Oil, or other fat things in close crucibles, giving a REGULUS like to the Antimonial; and these REGULI make Tin hard, to sound and be compact; if to one pound one ounce be added in fusion, for making good household stuff. And in trial they give good Gold.

And as it is said of purging Antimony, so also it is to be understood of the rest, as WISMUTH, ZINCK, LAPIS CALAMINARIS, Lead, Tin, Iron and Copper, to be purged from their superfluous sulphur, if you will draw more perfect metals, viz. Gold and Silver out of them with gain. And so I make an end of metallick lotions; recommending to Chymists, NITRE, TARTAR, FLINTS, AND LEAD; for who knows to use them, shall not Lose his labor in Chymistry: but this is to be lamented, that everywhere good earth and fixed in the fire, is not to be gotten, retaining Lead and Salts; for without our old Saturn little or nothing can be done in refining metals; therefore who goes to try anything in this Art, let him seek the best earth retaining Lead; twenty four hours space afterward let him consult with Tin, what VULCAN has to be done with Iron; who will tell him what he must suffer, before he obtain the Crown.

---

[20] Possibly Orpiment: arsenic trisulphide. -PNW

Of the Tincture of Sol and Antimony.

Sometimes an alteration happens to man's body, from the attraction of mineral vapors (which cannot be done by my Furnace) in the trial; therefore, here I will set down a certain medicine for the Workman's sake, as well for preserving as curing, namely, a clear rubin fixed, and soluble of Gold and Antimony. Take of pure Gold half an ounce, dissolve it in AQUA REGIA; precipitate the solution with liquor of Flints, as before is said in the Second part; edulcorate and dry the calx, and it will be prepared; take REGULUS MARTIS (of which is spoken in a little before) beaten fine, to which mix three parts of the purest Nitre; place the mixture in the crucible between burning coals, putting to fire by degrees: which done make a stronger, viz. for fusion; for then the mass will be made purple, which taken forth and cooled grind very small, of which take three or four parts and mix with one part of the aforesaid golden calx; place it mixed in a strong crucible covered over in the aforesaid wind Furnace, and make the mass to flow together like a metal, and it will assume the Antimonial Nitre in the fusion, and will dissolve the Gold or the calx of Gold, and a mass of an Amethyst color will be made therewith, which so long leave in the fire, till it gets the clearness of a Ruby, which one may try with a clean wire or iron bowed and put therein, although in the meantime the mass deprived of fusibility, is thickened; it is meat to add some nitre or Tartar, for speeding fusion, and that as often as shall be needful. Lastly, pour the mass, when it shall come to the utmost redness of a Ruby, hot into a clean copper mortar, which there leave until it cool, and it will be in color very like to an Oriental Ruby; then bruise it hot into powder, for taking air it would melt, and extract the tincture by the affusion of the spirit of Wine in a Vial, and the Gold together with the Antimony will remain very white like the finest Talc, to be washed with clear water, in a glass, edulcorated and dried; which melted with a stronger fire, gives a Yellow glass, in which no Gold appears, yet separable by way of precipitation with the filings of iron and copper, from which it recovers its ancient color, but without profit, by reason of the waste, the tinged spirit is to be taken away from the tincture, which is a very sovereign medicine in many grievous diseases.

    Although you may suspect this not to be the simple tincture of SOL, but of Nitre and Tartar mixt, be sure that the quantity of Nitre added not to exceed; and suppose that tincture of Tartar and Nitre, I pray what waste is there? Since that is so good a medicine by itself, & I am persuaded, this tincture of SOL to be better than those set down in the Second part. That Ruby may be so used by itself with proper vehicles, seeing it is a sovereign medicine of itself; or else exposed to the air and resolved to a liquor; for the medicine is no less than a tincture, because the Gold in it, and the purer part of Antimony are made potable without corrosives. Wonderful is the power of salts in metals to be destroyed, perfected and changed by fusion; for it happened to me one time making this Ruby, placing two other crucibles also with metals, by this containing gold with the prepared REGULUS of Antimony (for easily two or three, or more crucibles may be placed in this furnace, to be ruled with one fire, which cannot be done in a common furnace by that means) about to put in a certain salt into the crucible next to the crucible of gold, that by a mistake I cast it into the crucible with gold only, whence so great a conflict arose, that there was danger of boiling over; therefore forced to remove it out of the furnace presently with tongs, and to effuse it, supposing that the Ruby was lost by my rash putting in of salt; therefore I would only save the gold. And I found the effused mass red like blood, purer than a Ruby, but no Gold; but white grains like Lead dispersed here and there in the salts, because of their smallness, not separable but by the solution of the salts, which being separated by the solution of water from the red tincture like blood, remained in the bottom of the glass, which afterward for fusions sake I placed in a new crucible in that furnace, but willing to try the fusion. I found the crucible empty, and all the Gold vanished, a little excepted sticking on the top to the crucible and the cover, which I took away and melted for experience sake in a new close crucible, but all of it presently feeling heat flew away like Arsenick, no sign being left in the crucible; and so I was deprived of my Gold.

At length I took the red solution, and abstracted the water from the salts, and I found the salt red like blood, which I put in a clean crucible in the furnace for to try whether any metallick body might thence be extracted; but I found the effesed salt deprived of all tincture and redness, which seems strange to me even to this day, that by help of this salt the whole substance of gold, viz. the tincture together with the remainder flew away, having so great a volatility.

Which labor afterward I would reiterate, but it happened not so at all as at the first time; there was indeed some alteration of the gold made, but its volatilization was not so great, the cause of which things, I think was the ignorance of the weight of the foresaid salt, cast in at the first time against my will.

And two reasons chiefly moved me to insert this history: First, that it may appear how soon one mistake in a small thing frustrates the whole process. Second, That the truth of the Philosophers may appear writing that gold by art is reducible into a lower degree, equal to lead (which happened to me in this work) and that it is harder to destroy gold and make it like to an Imperfect metal, than to transmute an imperfect metal into gold; therefore I am glad in my heart that I saw such an experiment; of which thing our phantastick Philosophers will hear nothing, writing whole volumes against the truth, stiffly affirming, gold to be incorruptible, which is an errant lie; for I can show the contrary (if need be) many ways. I wonder indeed what moves such men to slight a thing unknown; I do not judge things unknown to me.

How dare they deny the transmutation of metals, knowing not how to use coals and tongs? Truly I confess those rude and circumforanteous Montebanks, not a little to defile and disgrace true Chymistry, everywhere cheating men by their fraud, being needy and oppressed with penury; unless per-adventure they find some credulous rich man giving them food and raiment for the conceived hope of Gain and Skill, of which also some being furnished with gold, go clad like painted Parrots, whom I judge to be hated worse than a Dog or a Snake, but innocent Chymistry is not therefore to be despised. Some covetous men besotted with folly and madness, laying out their moneys with an uncertain hope of gain, who afterward the thing ill succeeding, are forced to live in poverty, whose case is not to be pitied, destroying their money out of covetousness. Some seek wealth not out of covetousness, but rather that they may have wherewith to live, and may search nature, which are to be excused if they are deceived by knaves, yet not to be praised if they spend above their ability.

Another Tincture and Medicine of Gold.

Dissolve gold in AQUA REGIA; being dissolved, precipitate it with liquor of the salt of flints, pour some more of the aforesaid liquor to the precipitated gold, then place them in sand to boil for some hours space, and the liquor of flints will extract the tincture of the gold, and be dyed with a purple color; to which, pour rain water, and make it to boil together with that purple liquor, and the flint will be precipitated, the tincture of an excellent color with the salt of Tartar left; from which it is necessary to extract the water even to dryness, and a very fine salt of a purple color will remain in the bottom of the glass, out of which with the spirit of wine, may be drawn a tincture as red as blood, little inferior in virtue to potable gold; for many things lie hid in the purple salt, of which more things might be spoken if occasion permitted; therefore let it suffice to show the way of destroying gold, for that golden salt may in a very short time, viz. an hour, be perfected with small labor and transmuted into a wonder of nature; confuting the slanders of the noble Art of Alchemy; for which gift we ought to give immortal thanks to the immortal God.

Of Looking-glasses.

I have made mention in the treatise of **AURUM POTABILE**, not only of the material heat of fire, but also of turning the finest beams of the Sun into a material bodily substance, by help of certain instruments by which they are collected. I have also mentioned there a concave Glass, whose preparation I will here give, it being not known to all men, the best that I know is as follows. First, patterns are to be made of the best matter, namely, hair and Potters clay, of which thing in the Fifth part, conformable to the glasses, in form and figure circularly round; for else they cannot gather the Sun-beams together, and again put them forth; the fault of which thing is to be ascribed only to the pattern or mold: for the fusion and polishing of glasses is no singular Art, being known even to Bell-founders, but to melt them when very well shaped of the best matter and rightly to polish them, this is Art: and first to cut the patterns round, being very well Shaped by the use of a sharp Iron Instrument cannot briefly be demonstrated; therefore I will send the Reader to Authors prolixity handling this thing, viz. ARCHIMEDES and JOHAN BAPTIST, PORTA, and others; but if you want those Authors, or do not understand them, see you have a Globe exactly turned for making the Molds as follows: first make a mixture of meal and sifted ashes, which spread equally between two boards, as the manner is to spread past made of Flower and Butter for Pies and Tarts, answering in thickness to the glass to be shaped, then with a Compass make a circle as big as you please, which cut with a knife, and put it on the Globe, and sprinkle quick lime on it out of a searce[21] or five, and put clay well prepared with hair over it of the thickness of two fingers breadth, and if it be a great piece you must impose cross wires strengthening the Mold, least it be bent or broken. Afterward one part being hardened with the heat of the Sun or fire, take away all that from the Globe, and put it on some hollow thing, on which it may on all sides stand well, and also sprinkle quick lime or the powder of coals on the other side, and put upon this the other-part of the pattern, and again expose it by degrees, to be dried by the heat of Sun or fire, lest it crack; which done, take away the ends making those parts of the Mold or pattern from the inward or middle, which ends set one against another to the inward parts, the distance at least of a hands breadth, and put between in the top a few live coals to harden the Mold all over; to which put on other coals, and then more, and so by degrees even to the top, that they may be well kindled in their lighter parts; but if the Molds are very thick, one fire will not suffice, but it will be necessary to add more coals, until they be thoroughly kindled in the inner parts; afterwards, let the fire go out by degrees, that the types may grow cold, but not altogether, but so that you may touch them; and presently besmear finely the sifted ashes mixt with water, with a pencil, to stop up the chincks arisen from the burning the hair, and for smoothing the types; then again make both parts (after you have framed a hole in them for a Tunnel) clean, being wary lest any foul thing fall upon them; and carefully bind them together with iron or copper wire; and very well lute over the joining with clay prepared with hair; and put on an earthen Tunnel, and place the Mold in dry sand up to the top: And you ought in the mean while you burn and

---

[21] Sift. -PNW

prepare the Mold, to melt the metallick mixture, that it may be poured into the hot Mold, the Metal being well melted, cast in a bit of searcloth, which burning, pour out the melted Metal into the hot Mold, being wary lest coals or some other thing fall into the crucible, and be poured with the Metal into the Mold, spoiling the glass; then the glass cool of itself in. the Mold, if the matter does not moulder in the cooling: And if it should moulder in the cooling, which indeed would lessen it, it behooves that the cast glass be presently taken out of the Mold, and covered over with a hot earthen or iron vessel, that it may cool under it, which otherwise, Cooling shut up in the Mold not being able to moulder, is broke in pieces, but a little below you shall perceive, what be those mouldering metals.

And this is the common way (and the best) of melting, if you are exercised in your art: there are also other ways; first, when molds are made of wood or lead, agreeing to the glass, to be impressed with sand, or the finest powder of tyles or other earth, as is the custom of coppersmiths; and this way only serves for lesser glasses.

The third way which is the best of all, but hardest to one not exercised, is as follows; make a waxen mold with a Cylinder to be placed between two boards, as is aforesaid of the first way, which put upon the globe for to shape it, and let it be hardened in the cold; then take it away, and spread over it the following mixture with a pencil; which see that it be dried in the shadow, then apply potters clay, prepare with hair, the thickness of one or two fingers breadth; then take away the wax in manner following from the earth: make a round hole in the earthen mold with a knife, coming even to the wax; which done, place it near a coal fire, the mold being bending down, and the melted wax will run through the hole, into which pour the hot (not burnt) metal, & etc. that liniment which is anointed on the wax must be very well prepared least while the wax melt, it fall and melt away with the wax, nor let the wax pierce the earthen mold and spoil it. Now the liniment follows: Burn potters clay well washed in a furnace even to redness; afterward grind it and take away its finest part with washing of water, so that you may have an impalpable powder, which dry, and again burn with a strong fire: after grind it with rain water and salt Armoniack sublimed, upon a stone, as Painters use to prepare their colors, bring it to the just consistence of a paint, and mixture will be made; the salt Armoniack keeps that fine powder, lest it melt away with the wax: and the prepared earth makes a tender and fine fusion.

The Metallick Mixture for the Matter of the Looking-Glass.

There are divers of these mixtures, of which one is always better than the other, which by how much 'tis the harder, by so much the glass is the better; and by how much the harder the metal is, by so much the better it is polished; nor does the hardness of the mixture suffice, but its whiteness is also required: for red proceeds from too much copper; black from too much iron, or duskie from too much tin, and does not make the true representations of things, but changes the shape and color of them: for example, too much copper renders the Species redder than they are to be, and so of the rest; let therefore the metallick mixture be very white; but if burning glasses are to be made, it is no matter what color it be of, if so be that the mixture be hard. I will set down one of the best, Rx. of Copper plates the thinnest beaten to pieces one part, of white Arsenick a quarter part; first moisten the plates with the liquor of the salt of Tartar, and make a Stratum super Stratum, with plates and Arsenick powdered, by sprinkling this on them, until the crucible be filled; to which pour the oil of Linseed, as much as suffices to cover the copper and Arsenick; which done put on the cover with the best lute, then place the crucible (the lute being dried) in sand, so that only the upper part of the cover may stick out and administer fire by degrees, at first little; secondly somewhat stronger, till at length it be hot, that all the oil may evaporate; in the meantime, the oil will prepare the copper, and retain the Arsenick, and will make it enter into the plates, like oil piercing dry Leather; Or place the crucible upon a grate and put Fire to it, which administer by degrees, until the oil evaporate in the boiling. Lastly, when it shall cool, break the crucible, and you shalt find the copper of diverse colors, especially if you shalt take Orpin instead of Arsenick, and twice or thrice increased in magnitude, and brittle.

R. of this copper one part, and of latton (ORICHALCUM) two parts, melt it with a very quick Fire, and first indeed the latton, to which afterward add the friable copper; pour out the mixture melted and you shalt have a very hard metal unfileable, yet not so brittle, but like steel, of which diverse things may be formed serving instead of iron and steel instruments, take of this hard metal three parts of the best tin without lead one part, melt and effuse it, and the matter of looking glasses will be made. This mixture is a hard-white metal making the best-looking glasses, but if this labor seems tedious, take of copper three parts, of tin one part, of white Arsenick half a part for the matter of looking—glasses, which are fine but brittle, as well in the melting as polishing, therefore carefully to be handled. I must here set down a thing worthy to be observed, and known to few; viz. a false Opinion of many, especially of those who attribute knowledge to themselves of the properties of metals. In the second part (of subtle spirits) mention is made of the pores of metals, for experience witnesses, that those subtle spirits as of harts—horn, tartar, soot, and sometimes those sulphureous ones of salts and metals do evaporate through pewter vessels, which at the first hearing every man cannot conceive, for whose sake this discourse is made. Make two balls of Copper, and two of pure Tin not mixt with lead, of one and the same form and quantity, the weight of which balls observe exactly, which done, again melt the aforesaid balls or bullets into one, and first the copper, to which melted add the Tin, lest much Tin evaporate in the melting, & presently pour out the mixture melted into the mold of the first balls, and there will not come forth four nor scarce three balls, the weight of the four balls being reserved; if then metals are not porous, whence I pray does that great alteration of quantity proceed? Therefore, know that metals are porous, more or less; gold has the fewest pores, silver has more, Mercury more than that, Lead more than Mercury, Copper more than Lead, and Iron than Copper, but tin has most of all.

  If we could destroy metals, and again educe them destroyed from power to act, surely they would not be so porous. And as a child without correction is unapt to any goodness, but corrected is endued with all kinds of virtue and learning, so also we must understand of metals which left in their natural state, namely drawn out of the earth without correction and emendation remain volatile, but corrupted and regenerated are made more noble, even as our bodies destroyed and corrupted, at length shall arise clarified before they come into Gods sight. Well said PARACELSUS, that if in one hour metals were destroyed a hundred times, yet they could not be without a body, reassuming a new species and indeed a better, for it is rightly said, UNIUS CORRUPTIO, ALTERIUS GENERATIO; for the mortification of a superfluous sulphureous body is the regeneration of the mercurial soul, for without a destruction of metals perfection cannot be; therefore metals are to be destroyed and made formless, that thereby the superfluous earthy combustible sulphur being separated, the pure fine Mercurial species may spring forth. Of which thing more, when we speak of Artificial Stones.

Of the Smoothing and Polishing of Looking-glasses.

A Looking-glass though it be very exactly melted and proportioned, yet it is of no value if not rightly polished and smoothed; for easily in the smoothing any part it may suffer some damage hurtful to it, and it is necessary to take from them first, the grosser part by the wheel, as the custom is with Pewterers and Copper-smiths with a sandy stone, then to apply to them a finer stone with water, until they are sufficiently smoothed by grinding; which done, the looking glasses are again to be taken from the wheel and to be moved to the small wooden wheel covered with leather, rubbed over with a fine prepared glazing stone until the crevasses contracted in the turning no more appear, having got a cross line, afterward another small wheel covered with leather is required, to which a blood-stone prepared and washed with the ashes of tin rubbed on, to which likewise by the aforesaid means, according to the same line, the looking-glasses are so long to be moved till they get a sufficient fineness and brightness. You must keep such looking-glasses from the moist air, and breathing, and to wipe them when infected with air and breathing not with any woolen or linen cloth, but with a Goats or Harts skin, and not any way, but according to the cross line, and with which the looking-glasses are smoothed. They may also be smoothed by lead artificially melted, by first rubbing them with a smiris and water, and then with a finer smiris and lead; lastly with a blood stone and ashes of tin; likewise, also with whetstones, by changing for a finer every time, whence at length also they acquire a splendor by the ashes of tin.

Also, the outward part of the looking-glasses (convex) may be smoothed, which represents the species short, and spread the dispersed rays: but the inward part (hollow) gathers and multiplies, and puts forth or exposes the image.

Let these things suffice concerning the melting of looking glasses, & polishing requisites, for the collection of the Sun beams, and although from the aforesaid mixture other kinds of looking-glasses might be made representing wonderful shapes and several excellent things, as Cylindrick, Pyramidal, Parabolick, & etc. they are omitted as impertinent to this place, yet I could show a way to make them, because I have undergone no small labors and charges in the searching of their preparation and use, if it were necessary. But of all looking-glasses that is most useful whose preparation we have shown, whose diameter is at least two or three spans, if you will perform any special thing; although it be but of one or two spans, yet it gathers abundance of beams, so that you may melt tin and lead with it, if it be well shaped: yet the larger are the better. Nor ought they to be too deep, that they may cast their beams the further, and better perform their actions of functions, let them have the twentieth or thirtieth part of the sphere (the section being exactly observed) which is the foundation of the Art.

Of Artificial Gems, and Metallick Glasses.

As for metallical glasses pertaining to Alchemy, and much conducing to the perfection of metals, and esteemed by the Ancient Philosophers, I would not omit to say somewhat in this place, because they are easily made by this furnace.

And indeed, the Ancients have found these glasses questionless by chance, in reducing the calcined bodies into glass by a strong fire, for very many secrets by this means not sought for are found out. Oftentimes it happens to our labors, that past hope we find something better or worse, than the thing sought; and I think it has thus happened with these glasses, but however it be, I am sure these glasses have stood us in much stead; for ISAAC HOLLAND says plainly, that vitrified metals, being again brought to metals, by that reduction do give better and nobler metals than the first vitrified; and indeed gold gives a tincture, but silver gold, and copper silver; and so consequently the glass of other metals give better metals in reduction, the truth of which experience proves, and although I have not yet made great trial in this work, yet I know that metals brought into dead ashes to be turned into clear glass cannot be again reduced into metals without great profit: yet one metal is more pliable than another, nor are our glasses the Artificial stones of gold-smiths fixed to other large ones for ornaments sake, made by the addition of glass made of fusile sand; but ours are made of the juice of metals. But I do not deny the virtue of Venice glass, and others in the mundifying of metals, chiefly copper and tin, which yet is not comparable with metallick juices. I freely confess I have tried this thing twenty times, and I never was deceived by it. But I know not whether it may prove so in a greater quantity, because I never tried it, doubting of my vessels not fit to retain fusible glasses a requisite time: for I have spent much labor in making these kind of vessels, but hitherto in vain. For there is very great hope of gain, if you have very strong crucibles, nor is this perfection of metals without reason, for whilst the metal is burnt to ashes, much of the superfluous combustible sulphur is burnt (as you may see in Lead, Tin, and Copper, from the sparks appearing in their calcination whilst they are stirred and separated) which if again reduced (viz. being calcined) its better and heaver part (by benefit of melting) sinks to the bottom, the worse flowing on the top is changed into dross or glass. And so, the separation of metals is made by the help of the Fire alone, to the ignorant and inexpert incredible: but consider gilt silver to be separated in fusion, which is as it were corrupted by the common sulphur, and the metallick species, being lost, it turns to a black dross before that in melting it forsakes the gold: which way also silver is separated from copper, and this from iron. Observe also that black and crude Antimony, being reduced into ashes by calcination, and melted is separated by a strong Fire, the purer parts descending pure and white like silver, but the impure parts ascending are changed into glass or dross, which separation would never be made without incineration although the Antimony should have stood long influx.

You see therefore the power of Fire alone in melting metals, wherefore believe you that your labor shall not be in vain if you know how to help the Fire. Exercise yourself therefore in it, for you are sufficiently instructed, and this furnace will help you; without which it is impossible to manage such things well, as experiment testifies, confirming my words.

Mention being made of metallick glasses, which belongs to the perfection of metals, I am forced to say something also of other AMANSA, or colored glasses, which are called Gems, and are worn for beautifying, which though it be not profitable, yet it is a delightful labor, which knowledge, as well noble as ignoble have long sought, not for gain, but recreation sake, erring from the true way (although prolixity described in many tongues) through ignorance of the art to render crystal or flint fusible, and coloring it, being content with lead glasses made of one part of crystals, or flints, and three or four parts of minium or ceruse, glass of no worth, as not only very soft and unapt for polishing, but also heavier than it ought by means of the lead, and having a yellow or green color, for every glass made of crystal or flint, and minium or ceruse by themselves, viz. without the addition of other colors, gets a yellow color from the Lead, hindering and altering other mixt colors; therefore a good stone is not made this way of lead and flint, but Leaden glasses of this sort, Venice glass, Ashes of tin, and colors being added to them, be used diversely of the gold-smiths, namely to color gold, otherwise of no moment.

Therefore I will give another preparation, namely out of flints and crystals alone without minium and ceruse, with metallick colors, having the color and elegancy of excellent stones; but not harder than glass; for although crystal is harder than iron, yet by melting it is deprived of its hardness in some measure, and is made like to glass, yet so much hardness reserved, as serves to write on another glass, which glasses are easily polished, and in all things and by all, most like, hardness excepted, to natural stones; with which not only various kinds of stones may be made, and other gold, silver, and wooden works or pictures adorned; but also diverse supellectils[22], as salts, hasts or hilts, cups, & etc. and also images and antiquities may be formed (by fusion) like to those cut out of gems by the hand of an Ingenious workman, most delightful.

They are made after this manner: first you must look for flints and crystals not colored, but very white, gathered out of sand or streams, which you must heat in a covered crucible, and quench them glowing hot in cold water, that they may crack and may be pulverized; otherwise they are so hard that when they are powdered, they take part of the mortar and so are defiled; therefore it is worth your labor to handle them well. Afterward Rx. of flints, prepared, and the purest salt of Tartar, made in glazed vessels, but not in copper or iron, equal parts, mingle them and keep them for use.

---

[22] Spelling verified. -PNW

And if you will made this mass into a gem, you must first mingle some color (what you desire) afterward so long place it (being put into a clean covered crucible scarce half full) in a very strong fire, till all the salt of Tartar has evaporated, and the flint together with the color come into substance fusible like glass: you must then put a small clean iron wire, and draw out a little of the melted mass for trial; whether it have stood long enough in the fire, whether there be yet pustles and little sands, or whether it being exactly melted, it shall descend to the bottom, which done, you must take off the crucible, and place it under some hot iron or earthen vessel, that it may wax cold with the melted stone; otherwise the mass will be broken in the crucible into very small parts, and would be unfit for greater works: neither must you pour out the melted mass for fear of the attraction of Aire, and pustles to arise thence. But being willing to make out of the Mass by Fusion, not Engraving Money or Images; there is no need to leave the mass in the crucible to cool, but presently to pour it out hot in a copper mortar, and nothing will stick to the crucible, but all the mass will be poured out without any waste: And this mass, if you will, you may powder or break into very small bits for fusion and impression. But the mass when cooled in the crucible, is to be taken by breaking the crucible, and to be reduced into greater or lesser stones by cutting; but melting for money or images; you must place the money or image, which you will imitate, with the backside or hinder-part downward in an iron Ring, a Fingers breadth broad of greater capacity than the money, upon a stone or plain wood, and sprinkle on a little Tripoly, or fine Sand, through a cloth, namely, as much as suffices to cover the mold, and upon this to put more, well moistened with water, like ashes of cupels, and to press it, being most tenacious, firmly to the mold, but warily, lest the mold move; which done, you must turn the ring, and with a knife lift up the mold, and to take it, being lifted up with ones hands or tongs, the image being left in the sand, to be dried by heat of the Sun or Fire. Afterward to cast the image, place the ring with the image impressed in the sand under a tile, and administer a strong fire, that the whole ring, with the sand, and the image in the sand may be very hot: then take off the ring, to see if the image has suffered any loss; which, if it have not, you must put upon it so much of the aforesaid glass, coarsely beaten, as suffices in the fusion to fill the image impressed on the sand; which done, put the ring again under the tile, and administer a fire of fusion, till the glass melt in the ring; to which, touch with a smooth iron and light, (with a handle) being hot the ring being taken out of the furnace with tongs, pressing the glass well to the mold; and then place it under a hot iron, or earthen vessel to cool; and being cold, take the image from the mold, which answers to it in all things, if you have correctly proceeded, exactly representing the Carvers art, or a seal impressed on a jewel, which excellent work is most fit to feign, and represent Antiquities and Rarities.

The coloring of the aforesaid mass follows, by which it is made most like to Gems.

It behooves that colors be taken from metals and minerals, namely from Copper, Iron, Gold, Silver, Wismuth, Magnesia and Granate; of other colors I know nothing of certainty, Copper commonly makes a color green like the Sea, Copper with Iron, grass-green; Granate a smaragdine color, Iron yellow or jacynth; Gold the best skie color; Wismuth common skie color; Magnesia Amethystine, mixt, they give other colors; E. gr. Gold mixt with Silver gives an Amethyst color; Iron and Copper, a pale green; Wismuth and Magnesia, a purple; Silver and Magnesia, various colors like an Opal.

Images are also made of divers colors, if the masses of diverse colors be broken into bits and mixt, be put upon the Mold, & etc.

And if you desire an opac mass (green, red, skie color, & etc.) add a little calx of Tin darkening, on which as on a Basis the colors insist. For example, in making a Turquoise stone or a Lazulus, mingle with the Azure made of the silver Marcasit or Zafora, (to the color of the mass) the calx of Tin, that they may melt together, and before the impression be made, put upon the Mold some prepared gold, then spread and put upon this the aforesaid glass; and the fusion and impression being made, will be made thence a stone having golden veins like LAPIS LAZULUS very delightful; But there must be a calx of Gold not losing its splendor in the fire, such as is made by Mercury, or that which is better, which is precipitated out of AQUA REGIA: of which above.

Of the preparation of the colors for coloring the mass of Flints and Crystals.

The plates of copper often heated, are to be quenched in cold water of which more in the Fifth part, from three to six grains of it may be mixed with 1 ounce of the mass for a Sea-green color. Iron reduced into crocus by reverberation; of which from four to 10 grains are added to the mass for a yellow or Jacynth color; Silver is dissolved in AQUA FORTIS, and precipitated with the liquor of Flints after it is edulcorated and dried, whereof from one to six grains, added to 1 ounce of the mass, they make mixt colors.

Gold is dissolved in AQUA REGIA, edulcorated and dried, precipitated first with liquor of Flints, whereof from four grains to 1/24 ounce mixt with one ounce of the mass, make a most elegant Sapphire. And if from three to six of that soluble ruby made of the Gold, and the nitrous REGULUS MARTIS be added to 1 ounce of the mass, they make a very polite ruby: Magnesia pulverized, whereof, from six to fourteen grains, to 1 ounce of the mass, makes an Amethyst.

Marcasite dissolved in AQUA REGIA precipitated with the liquor of flints, edulcorated and dried, whereof from one to five grains, to 1 ounce of the mass, give a Sapphire, but not comparably so polite as one made with gold.

But being unwilling to calcine Marcasite, let him take Zafora, and mingle to 1 ounce from five to ten grains; Granates of BOHEMIA, or Oriental pulverized, add from six grains to 1/24 ounce to 1/8 ounce of the mass, for little green stones like to the natural smaraged or emerald: other things which remain of the mixture of the colors, are to be learned by experience.

To what uses colored flints and crystals are appointed, is not here to be treated of; one use excepted, which I set down for the eyes, which are weakened by too much watching, the heat of fire and smoke; see you have a waxen mold circularly round, of the bigness of a dish or trencher; (the Optiques are wont to call such LENTES) to which, put the best clay well mixed with hair; anoint the waxen type with oil, and exactly apply the best prepared earth of crucibles (and durable in the fire) the thickness of a finger; which being dried, perforate in some part, that the wax being melted by the fire, may flow forth: afterward burn the mold in an earthen furnace; being burnt, fill it with prepared glass, and place it in a wind furnace till the glass melt; which at length being cooled, take off the type by attrition, and there shall you have the crystal resembling the form of the type; which afterward you must make and polish like spectacles in an iron dish on both sides; and take it out with a strong iron wire, and you shall have a good crystalline LENS for a small-price, which otherwise is scarce made of crystal of so great a bigness. And if you will, you may color the glass green, very pleasant to the sight, and fit a foot to it for greater benefit. And the glass does not only serve for the Multiplication of light in the night time, that you may see a thing far off in a chamber, but also for the fixing and calcining minerals by the Sun-beams, and melting of Metals, and multiplying of Pictures, like a hollow glass, and also for other uses it may be compared with a hollow looking-glass, which does the same of an equal bigness with the hollow glass; nor is there any other difference of them but reflection. This glass instrument is made likewise another way, and by less cost and labor, if it be of a polished looking-glass, if two great orbs are cut out with a diamond, and if they are somewhat softened with fire, and are left there so long in the heat, until they shall stick like wax very close to the stone, which done, let them be cooled again, which afterward taken out, will represent the form of a hollow glass; to which, it behooves to fix a leaf on the convex part. And the glasses do the same than a hollow metallick looking-glass does, the reflection excepted, which is not so strong as of the hollow glass: And although the glasses are sooner broke; yet they are very fit for the making of the following Instrument.

And they are bound together with a strong wire, applied across on the concave part, and ash hole is cut in the brim with a diamond on one side, of the bigness of a pea, then the crevasses are exactly closed in every place with the best lute; which done, a silver or copper ring is to be tied about it, holding those glasses straightly, so that the Instrument may be fitted to the foot, all which well done, those strong wires are separated or cut off, with which the glasses were bound at first, namely, near the copper ring: afterward very pure AQUA VITA is to be put in through a funnel, as much as is required for the filling it up, the Instrument being filled, the hole is shut up, which is to be kept for use; and this Instrument is better than the hollow glass; especially if it has its diameter the breadth of one foot, and may be applied to prospective pictures, it does excellently represent and multiply them.

Behind which, if you place a candle in the night, it gives so much light in the Chamber, that you would think it came from the Sun. It does also many other things which are here omitted as superfluous. And you may gather the dispersed light in the Aire in the night time with it, so that you may read the smallest writing. Such and others of the like things may be done by this furnace, all which to set down, would swell the Book too much. Other things of the metals examination and purification by fusion, in another place.

Take this, Reader, which is given to you, in good part, at another time you shalt have better; and do not mistake my writings, as if I did reprove the examinations of metals by the Ancients, fusions and separations, who only would communicate my opinion, and yield my assistance for further proceeding; for I know that dealers in metals giving too much credit to their small proof when they find nothing, do contemn ores as barren, often abounding with gold and silver; when nevertheless, JOHN MATHES says expressly in his SAREPTA, that minerals oftentimes tried in a small quantity do yield no gold and silver, which in a great quantity, yield a great deal, wherefore credit is not always to be given to such trials, often deceiving, as experience testifies.

And this not only in those minerals which are digged out of the earth; but also in those calyie and sandy minerals, abounding with silver and golden flames; out of which neither by the less nor greater proofs, nor ablution nor Mercury is drawn with gain, that thin and fiery dispersed gold: which by some waters is done without fire easily; for I know such mines are found near many rivers of Germany, and many places in other Nations of Europe, out of which honest gain without much cost and labor may easily be gotten. Neither are they dreams, which I have spoken parabolically of the perfection of metals, for it is possible by art to help nature in the perfecting things. There is therefore no more need of anything than of knowledge; therefore, the nature of metals being known, and their properties, they are easily separated, purged and perfected.

But what I have written of the universal medicine, I have done for the aforesaid causes, which have made me believe the thing, not as professor of the Art. The other things of colored red glasses and looking glasses I have added, because they are easily prepared by this furnace, as sometimes necessary in some works. Other things of the handling metals are not without cause now omitted, which happily may be sometime delivered in another place, wherefore now we end.

FINIS.

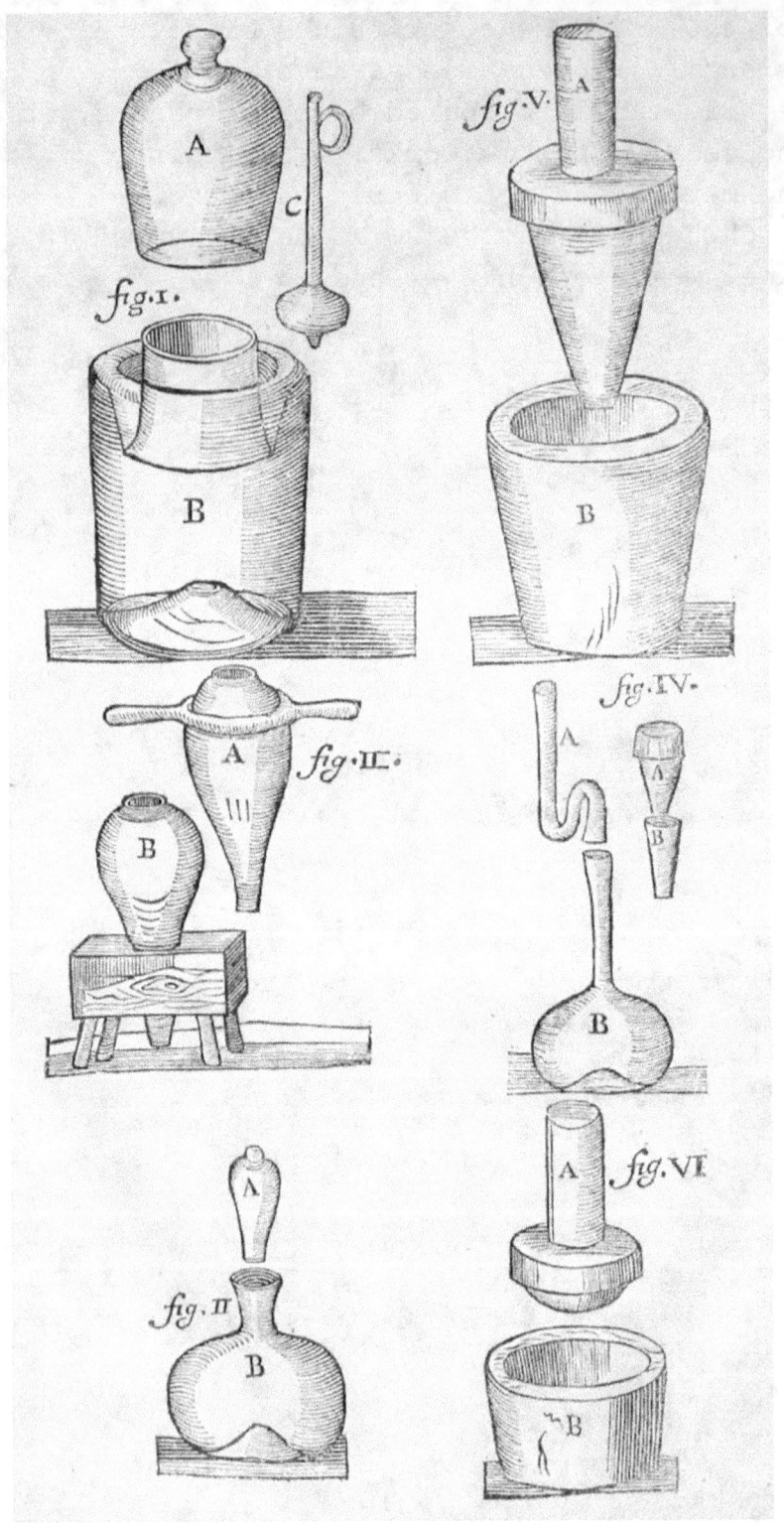

Fifth Part of Philosophical Furnaces

In which is treated of the wonderful Nature of the Fifth Furnace: Also, of the easy Preparation of the Instruments and Materials belonging to the foresaid Four Furnaces. Most profitable for Chymical Physicians.

Of the Preparation of the Furnace.

As concerning this, of which, though I made no mention in the Preface; for it was not my Resolution to mention it in the last Part, because I was purposed only to treat of the Instruments, as well earthen, as those of glass, and also of the other necessary things belonging to those four parts premised; yet I am willing now in this Part, (which I have judged to be the most convenient place for it, for which I did before design another) to discover the wonderful Nature thereof, as far as I may for the Studious Artist's sake. And although I know that more in this part, than in all my other writing's especially the ignorant and unskillful, will be offended; yet I will not therefore pass it by, persuading myself, that by this means I shall do a work, that will be most acceptable to the searchers of Art, and Nature. For I do devoutly affirm, that this is the choicest of all my secrets that I confide in, in which I have already seen wonderful things, hoping that the Divine Benediction will some time or other be obtained upon the practice thereof. And as for the structure of it, much cannot be said thereof, because it is not built as other Furnaces are, but it is everywhere found extructed by Nature, being ordained for no other works, than those of Nature, viz. for the making of any MENSTRUUM that shall dissolve gold, silver, and all other metals, and minerals without any noise, as also precious, and common stones, and also glasses: the original of which, is the original of the MENSTRUUM. Now what, and what manner of Furnace that is, that produces this Royal MENSTRUUM, (coming from the MENSTRUUM itself) and that easily without any labor, you may easily conjecture, that it is not any common one, by the help whereof other things are distilled, that can yield such a MENSTRUUM that is not corrosive: which certainly is not any common MENSTRUUM, because there is but this one MENSTRUUM that I know, which does not partake of any corrosive quality, that does more than any of all other corrosive waters whatsoever. For all corrosives, whatsoever they are, as AQUA FORTIS, AQUA REGIA, Spirit of salt, vitriol, allome, and nitre cannot together, and at once dissolve the close union of gold, and silver, and other most hard subjects that cannot be dissolved in waters, though ever so caustic.

    This indeed is wonderful, and stupendous, that a thing everywhere found most vile and base, should do so great a miracle: I know not what moved me to write of it, knowing that I shall in this part offend not only the wise by writing so openly, but also the ignorant detractors, and slanderers that will accuse me of falsity. And truly these considerations might justly have deterred me, but that I knew I might do a good work, recalling many from their errors. For many are persuaded that there is no other dissolving MENSTRUUM, besides the aforesaid corrosive spirits; but those are Chymists that are ignorant of Nature; yet the Philosophers with one consent say, that those corrosive destructive spirits make a fruitless solution of metals; for experience testifies that the solutions made by the help of AQUA FORTIS, and REGIA, and other spirits, color the hands, being that which a true Philosophical solution does not, and furthermore testifies that those, viz. which color the hands are not to be reckoned among the true Philosophical solutions, but to be contemned as Malignant. Wherefore I was willing to write these things to instruct those that err. Let no man therefore persuade himself, that a MENSTRUUM so vile and contemptible, is of less efficacy, than those corrosive spirits. I myself did once scarce believe, that such great Virtues, could be in so most vile a MENSTRUUM, until I had experience of the truth in good earnest.

 I could here add more things concerning the original of the universal MENSTRUUM, which is to contemptible, which does by its wonderful powers and virtues dissolve all metals, minerals and stones radically without any noise, unites and fixes them; the solution whereof does not color the hand; the conjunction is inseparable, and the fixation incombustible. I say, I could add more things concerning it, but that divers inconveniences, which by this means I might incur, as also the envy and hatred of others do deter me. For although anyone who does think to discover the possibility of Art, and Nature; yet few would be content therewith, being very desirous of all manner of revelation, and if we should not gratify them, we should forthwith incur their hatred and envy, who would without doubt judge otherwise of the matter, if they had but any experience of our labors. Be you therefore (courteous Reader) contented with this discourse, that shows you the possibility of Art and Nature; and diligently seek after it in the fear of God, and without doubt your labor shall not be in vain.

Of the Building of the Furnaces.

How those Furnaces of the first and second part are to be built and made of Potters Clay, and Stones; I need not say much, because there be many Books extant, treating of this matter sufficiently; yet this caution is to be observed in building of the Furnaces, viz. that those Furnaces, in which a very strong fire is not kindled, need not so strong walls, as those in which we distil, sublime, and melt, with a most strong fire. And for what belongs to subliming and distilling Furnaces; you may erect them of those common bricks which are made of the best clay, and well burnt, compassing them about with very strong walls, that they may the longer retain the heat: or else you will continually have something to do in mending them, and closing their chinks, which hinder the regiment of fire. Wherefore they must be compassed about with iron hoops, that they may be durable and not gape. Now what concerns the melting Furnaces, the aforesaid bricks are not of use in the building of them, because they not being durable melt in the fire; wherefore you must make other bricks of a very good earth that is fixed in the Fire, such as is that of crucibles, & etc. of which, afterwards; which are to be made in a brazen or wooden mould, and to be burnt, and it matters not whether they be round or square, a regard being had of the Furnace, that six or eight of them make one course, or row. But you need not build the whole Furnace of these stones, for it is sufficient, if the place only, where the coals still lye, be made of them, and the other part of the Furnace be made of common bricks.

A Lute for the Erecting of Furnaces.

Lute may be made divers ways for this business; for men prepare their Lute several ways as they please. Some mix with sifted Potters earth, the beaten hairs of Cows, Oxen, Harts, or the chaff of Barley, Tow, Flocks, Horse dung, and the like, that hold together the clay, and prevent chops, to which they add sometimes sifted sand, if the clay be too fat, beating the mixture together with water, and bringing it to a just consistence. And this is the best mixture, that is not subject to cleaving, yet weak, because in length of time the hair and chaff are burnt, wherefore the Furnace becomes thin and weak. Many leave out combustible things, and mix Potters clay, and sand together, and temper them with brine, for the making of their Furnaces. And this is the best mixture, because it is not combustible as the other is, neither is it subject to cracking, by reason of the salt: and for this purpose, the brine of fish and salt flesh so serve, and is very good, because the blood helps the joining of them together: but if the CAPUT MORTUUM of vitriol or AQUA FORTIS, being mollified, be mixed with Potters clay and sand, you go a better way to work: for this Lute is not at all subject to cracking, but fixed in the fire and permanent. With this Lute are Retorts, and Gourds very well luted, and coated, also the joints of Retorts, and Receivers closed: this being mollified with a wet cloth applied to it, may again be separated, and taken off, as that also with which salt is mixed: but the other Lutes that want salt will not be separated, by reason whereof glasses oftentimes are broken. Wherefore in defect of the CAPUT MORTUUM of Vitriol, temper the clay with sand and with brine. But many mix the filings of iron, powdered glass, flints, & etc., but you need not them for the building of the Furnaces, but only for the coating of certain glasses used for separation, and distillation, because the filings of iron being helped with salt, binds, and joins together most strongly.

Of the Closing of the Joints, Hindering the Evaporation of Subtle Spirits.

The aforesaid Lute is sufficient for the closing of the joints of the first Furnace, where air is not kept from the Spirits, but not of the Vessels of the second Furnace, where most subtle Spirits are distilled, which it cannot retain, penetrating the same with the loss of the better part: wherefore you must make choice of another; unless upon the other being well dried, a mixture made of quick Lime, most subtly powdered, and Linseed-oil, be smeared over with a pencil, which the porous clay attracting to it, is fortified, so as to be able to retain those most subtle Spirits: but this Lute can hardly be separated again; because refusing water, it cannot be mollified; wherefore the clay is to be tempered only with the white of eggs, and to be applied with linen clouts: but you must prevent the burning of the linen, by reason of the extreme heat of the neck of the Receiver, by putting between an iron or strong glass, viz. betwixt the receiver and the retort. The joints also may be closed with ox bladders wet in the white of eggs, also with starch tempered with water, if it be sometimes applied, being smeared on paper. For by this means those most subtle spirits are easily retained, but not corrosive, for which use the CAPUT MORTUUM of AQUA FORTIS is more convenient, which after it is dried must be smeared over with a mixture made of linseed oil, and quick lime.

And divers kinds of these lutes are had being destined to divers uses.

Another Lute for broken Glasses.

It happens sometimes that glass vessels, as receivers, and retorts, have some cracks, but otherwise are whole and sound; which are greater in those glasses that do again suffer the heat of the Fire, wherefore at last the glasses are broken, which if you will prevent, make a liniment or thin lute of linseed oil, quick lime, and red lead; which being smeared over a linen cloth apply to the crack, upon which being dried apply another: but if the crack be very great, you may apply three or four linen cloths, for the greater safety sake: as you may apply the whites of eggs beaten together, upon the cracks with linen, and cast upon it quick lime sifted very fine, and press it down hard with your hand: which being done, you may apply over them more linen cloths wet in the whites of eggs, and cast upon them quick lime again: which when the lute is well dried, retains the spirits, but sooner subject to the corrosion of corrosive spirits than the former.

Note well that quick lime is not to be mixed with the white of eggs, and so used upon linen cloths, as the manner of some is; because the whites of eggs acquire a hardness from the lime before they be united, and therefore cannot stick, but linen cloths wet First therewith before the quick lime be cast upon them, so that the lime does not immediately touch the glass, being applied betwixt two linen cloths.

How those Subtle Spirits when they are made, may be kept that they Evaporate Not.

Those glasses in which those spirits are kept are for the most part stopped with cork, or wax, upon which afterward bladders are bound: which stopping is convenient for some spirits, that do not prey upon cork or wax. For all corrosive spirits, as of vitriol, Allome, common salt, nitre, & etc. corrode cork; and lixivial spirits, as that of harts—horn, tartar, salt armoniack, urine, wine, & etc. melt wax, and penetrate it.

And although other stopples might be made, which might retain both sorts of spirits, yet it would be tedious and laborious to open those so often, and to stop them again. Wherefore I have found out a fit kind of glasses, viz. of such, whose mouths have distinctions, and are fit to receive their covers; as it appears by the delineation. (See the first figure[23]). A. signifies the cover: B. the glass containing the spirit. C. a drawer by the help whereof the spirits are taken out of the glass, when there is occasion, into the distinction in the brim of the mouth; viz. of the glass that contains the spirit, is put quicksilver, and upon this is put a cover; this being done, the Mercury closes the joints of both glasses running in the brim, so that nothing at all can evaporate; for the spirits do not penetrate the Mercury, unless they be very corrosive (a thing to be noted) which then in process of time

---

[23] These figures are located just before the start of Part 5.

turn the Mercury into water, but very seldom; and then the Mercury is to be renewed. But we need not give so much honor to corrosive spirits, being not to be compared to those volatile ones, which being abstracted from corrosives not prey upon Mercury; and much less than these, do lixivial spirits corrode Mercury; and for the sake of these were these glasses invented, by the help whereof most subtle spirits are without any loss of their virtues, if you please, a very long time preserved and kept. And because when there is occasion the spirits cannot be poured forth because of the Mercury in the brim, you must get a drawer like to that, by the help whereof wine is taken out of the vessel, but lesser, having a belly with a little mouth made very accurately. This being let down you may take up as much as you please, as are needful, the upper orifice whereof being stopped with the finger nothing drops out; being put into a lesser glass is thence poured forth for your use. Then you must again cover the remainder of the Spirit that is in the glass, and as oft as is needful take out with that drawer as much as is useful. And this is the best way by which the most subtle spirits are retained; which also are very well retained in those glasses, whose stopples are of glass smoothed with grinding. But this is a more costly way of keeping in spirits, and it is done after this manner.

How Glass Stopples are to be Smoothed by Grinding for the Retaining of Spirits in their Glass Vessels.

First of all, order the matter so that you have glass bottles of several sorts, some greater, some lesser, with strong necks, and mouths, with their glass stopples, which being smoothed by grinding shut the orifice of the bottle very close. Now they are smoothed thus: Put the stopples in the turn, being set or fastened in some wood, bring it into a round shape, then being moistened with SMIRIS and water mixed together, let it be put to the mouth of the bottle, so as to be turned round in the mouth of the bottle, which you must often take away from the stopples being fastened to the turn, for the more often it is moistened, which is with that mixture of prepared SMIRIS and water, with the help of a pencil, or feather; and that so often and so long, until the stopple stop the mouth of the bottle most closely: which being done, you wipe off the SMIRIS with a lint from the stopples and mouth of the bottle, then smear over the stopple with a liniment made of some fine washed earth, and water, or oil, and again turn it round in the mouth of the bottle, and often smear it over with this fresh mixture, until the stopple be most exactly smoothed, which afterward is to be tied to its proper bottle; the same also is to be understood concerning the rest; that one may not be taken for another, & etc. And that you may not need to take away so much from the stopples, and bottles, get some copper moulds made for the stopples, which stopples must be taken whilst they be yet warm, soft, and new drawn from the furnace, that they may be made of a just roundness, as also other copper moulds. Which must be put into the mouths of the bottles, whilst they be yet hot and soft, for the better making of them round, whereby afterwards the stopple may more easily, and quickly become fit to stop the mouths of the bottles very close, (as for example: A. is the stopple, B. the glass or bottle) if you know how to order them rightly, they will quickly and easily fit one the other.

In defect of turn, proceed after the following manner, which is slow, yet safe, because in a turn the glasses, oftentimes waxing hot are broken by reason of the over great haste; and it is thus, make an iron or wooden receptacle fit to receive the glass bottle, which being covered about with linen, and put in, join both parts of the receptacle warily and softly, with the help of a screw, that the bottle be not broken, and that that instrument, or receptacle of the bottle being fastened to a form with the help of the screw, cannot be moved. Afterwards cause that another wooden instrument be made for the stopple (as for example, A. the stopple with its receptacle, B. the bottle with its receptacle) that may be separated in the middle, and be again reunited with a screw after the putting in of the stopple, which being smeared over with the aforesaid mixture of SMIRIS and water, take the instrument with both hands, and put the stopple round about the neck of the bottle, and grind it round upon the other, as Wine Coopers are used to do in smoothing the taps, and that so long until the stopple be fit for the bottle; then reiterate the same labor with the earth TRIPOLIS, until it be completed; and it will stop as well as a stopple made by the help of a turn (See the second and third Figures before the fourth part).

After this manner also you must work those greater glass receivers of the first furnace, that without luting they may be closed. Stopples also of vials or Boltheads for fixation may be wrought after this manner, which instead of luting may be put into the mouths of the vials, upon which are put caps of lead; by which means in case of necessity they may be lifted up, viz. in case the spirits by too strong a fire be stirred up and rarified, by reason of the danger the glasses are in to be broken, and may again fall down into the mouths of the bottles being pressed down with the leaden caps, and so stop close again. And this way of stopping is better than that which is done with cork, wax, sulphur, and other things: because in case the fire be not well governed, and by consequence an error is committed, you may preserve your glasses by lifting up of the stopples, viz. when the spirits are too much stirred up. And although this be a better way of stopping than the other common way; yet that which follows is better than this, whereby the spirits are easily retained, the glasses being preserved, and without all danger of being broken. And it is thus, viz. get a glass pipe to be made crooked according to the figure set down, into the belly whereof is quicksilver to be put from half an ounce to an ounce, or thereabouts, and let this pipe which bath a belly be put into the vial containing the matter to be fixed (as for example. A. the pipe with a belly, B. is the vial, and again C. signifies the aforesaid leaden cap with the neck of the vial D.) the joints whereof afterwards are to be covered over with lute, and the vial will never be in danger of being broken. See the Fourth Figure.

    These foresaid ways of stopping are the best, by which the breakings of glasses are prevented, viz. whilst men are in error about the fixing of spirits of salts, minerals and metals, which although they are fixed with great costs and labors, yet do not satisfy what is promised and expected, because those kinds of fixations are violent and forced, and by consequence contrary to nature: but in the profitable fixation of spirits, not so, where we must follow Nature, and not commit ourselves to fortune in our labors. For only fools are wont to break their glasses in their supposed tincture; but Philosophers not so; for every VIOLENT THING IS AN ENEMY TO NATURE; and all the operations of Nature are spontaneous. They err therefore, and never shall come unto their desired end, who attempt violent fixations. I cannot be persuaded that bodies dead, or half dead can be so mixed as to multiply: but I could easily believe that the conjunction of male and female of one and the same species, sound and nourished with sound and wholesome meats to be natural, and to make a spontaneous propagation, and multiplication of their species; viz. of those that endure in a good, and adverse fortune, in life, and death; but the conjunction of dead things, to be dead, and barren. Do but consider how many and various instruments both gold, silver, copper, Iron, tin, and lead; as also earthen, glass, stone, and other vessels of other materials have been already invented, and found out for the fixing of Mercury alone with gold and silver, but in vain, because they have no mutual affinity. For although Mercury adheres to metals, or metals to it, yet that is not because of any affinity for multiplication, or perfection sake: for it appears by experience that Mercury flies away in the fire, and leaves the gold, silver and other metals. Where it is clear that they have no mutual affinity requisite for the multiplication of metals, nor is it ever possible: For they that have a mutual affinity embrace one the other and abide together forever, although volatile, yet never leave one the other, like gold and Mercury, when they are united together with the strongest bond, so that they can never be separated although with the strongest fire. Wherefore a great care is to be had in the fixation of things joined together; which if they have a mutual affinity, will embrace and retain one the other, without the help of any curious glasses with long necks. Of which things if you are ignorant, abstain from meddling with them, as being more hurtful then profitable, as daily experience both mine own, and others do witness. But that you may the better understand what things have a mutual affinity one with the other: attend a little to what I shall say.

Is not he to be laughed at for his folly who will pour rain, or common water on gold, silver, and other metals to fix them? See therefore the unwise actions of many covetous Alchemists in so hard a matter, that spend their time in trifles, reaping according to what they have sowed, and at last leave off their work which they have undertaken, after they have expended much cost, and spent their labor in stenches, watchings, and cares. For I have oftentimes seen those, that although they have not chosen common water for their MENSTRUUM, yet have made choice of MAY-DEW, snow or rain gathered in MARCH, and water distilled out of Nostick, or excrement of Stars; vegetables and animals for their solvent, in which they have lost their labor.

For as the radical union of the aforesaid things with metals is impossible: so never is any good to be produced from thence, because of their difference. And such may deservedly be compared to those, who ascending a very high ladder that has many steps, do presently endeavor to fly from the lowermost to the uppermost; which is a thing impossible; so, neither can there be any conjunction of things that do so much differ. But as any one may easily ascend the highest step by degrees, so also any one may (which yet he need not do) join together extremes, by adding first a thing that is most near to one of the extremes, and then to this another next to it, and so by consequence, until you come to the other extreme, which is a thing that requires a very long time, and is a work without profit. And if things be joined together that have the next affinity, the one will be delighted in the other, and the one will embrace the other, will overcome, and retain it. As for example, there is a certain salt, and that only, that can coagulate, and turn into a body like itself, even common water, which can be fixed in a very little time, with, and by one only certain mineral, which is very volatile. Minerals also may be fixed by metals, and metals, (a thing which I never tried) by a certain thing more excellent than metals, without all doubt. But therefore, it is needful in the fixation of minerals to begin with the coagulation of water, whereby it is turned into salt; and this afterward into a mineral; which would be too tedious; but it is sufficient to begin in things most near, in which nature has begun to operate, but has left Imperfect; for then there is hope of gain, if contrary things are not joined together, else not. Behold how ready Nature is at hand to help anything that is administered to it, which it can help: as for example, make salt of calcined Tartar by the help of solution and coagulation (but do not take that for it, of which a little before mention has been made, which is far better than salt of Tartar) of which after it is calcined, observe the weight; upon which afterwards pour half the weight of most pure rain water; distilled to avoid the suspicion of impurity; then draw off the water gently in BALNEO, or Sand, which again pour upon the remaining salt of Tartar, and again draw it off; this do so often as is needful, until all the water be consumed. Which being done, take out the salt, and weigh it, being first made red hot in the fire, and you shall find it to be increased in weight, which increase in weight came from the water, and not elsewhere.

Note well that the cohobation of the water is to be reiterated often upon the salt of Tartar. Observe, that by this means, the water is convertible into salt by Art, & etc. And If you do not believe the conversion of things material and corporeal, how will you believe the conversion of things immaterial, as the Sun, and Fire into a material fixed substance; of which thing, something shall be treated in our Treatise of AURUM POTABILE, and more at large afterwards in a Treatise DE GENARATIONE, if God permit: For you must know that the circulation of the Elements, and things elementated, viz. how one is converted into another; and how they nourish and cherish one the other: as for example, the Earth yields Water, the Water Air, the Air Fire, and the Fire again Earth; which if it be pure, yields pure Earth. But that you may understand aright how anything to be fixed, may be retained by another because of affinity, observe the following example. The Husbandman casting seed into the Earth for to multiply, does not choose any Earth, but that which is convenient for multiplication, viz. an Earth that is neither too dry, nor too moist; for the Seed cast in sand cannot grow, but is lost: For whatsoever is to be preserved, is to be preserved by an equal temper; which by how much it is more equal or like, so much the more perfect substance it does produce. Humidity therefore being necessarily requisite for the growth of vegetables, without which, they can neither grow, nor multiply, but the seed being cast into moist sand, and the Rays of the Sun acting upon the sand, and suddenly consuming the humidity thereof, whence follows the burning up of the seed in the dry sand, because there was no affinity betwixt the water, and Sand; without which, the water could not be retained by the sand, and consequently, the seed deprived of its nutriment; it follows necessarily, that some MEDIUM be required, or bond joining and binding the rain, and sand, viz. salt, by the help whereof, the rain water is retained by the sand, that it be not so easily consumed by the heat of the sun.

    The sand therefore retains the salt, and the salt, the rain water for the nutrition of the bud: but every salt is not convenient for this business; for although Christ says, LUKE CHAP. 14, Verse the last, that earth without salt is barren; yet any common salt is not to be understood thereby; (See more **DE NATURA SALIUM.**) for some salts, as common salt, salt of Vitriol, Allome, & etc. do not only not do good, but do hurt to Vegetables, hindering by reason of their dryness their growth and increase. Now lixivial salts promote them that which Country men do better understand, then our supposed Philosophers: for they know how to help their barren ground with the excrements of Animals; which are nothing else but a lixivial salt, mixed with sulphur, making the earth fat and fertile. And by this means a VEHICULUM (rather a bond) is administered to the rain water, that it may the less be consumed by the heat of the Sun. Moreover, all seed (consisting in a lixivial salt and sulphur) loves its like, from whence it borrows its nutriment, which is observed but by a few Learned or Unlearned. Husband-Men may well be excused of their ignorance, because they work only out of Use and Custom. But others that bear the Title of Learning not so; whose Duty is to render a reason of Germination, who may deservedly be Ashamed of their

Ignorance, being less knowing than Husband-Men. It is manifest, that Dung makes the Earth Fruitful; but how, and for what reason, not so; but if it did want nitrous salt, it would neither make it Fertile, nor promote Germination: for it is not unknown, that Nitre is made from the excrements of Animals. The goodness therefore of the dung consists only in the lixivial salt contained in it, and not the straw.

But you will ask perhaps, why does not any other salt help Germination? Why is the salt of dung required to Germination, and no other? We have already answered that, like are helped with like; and contraries are destroyed by contraries: For experience does testify, that every seed consists in a lixivial salt and sulphur, and not in any acid salt; wherefore also it does desire and embrace its like. Let him therefore, that will not believe it, make trial of the distillation of the seed of any vegetable; of which, let him force over a pound by a retort; and he shall see by experience, that not an acid spirit, but a flegme together with plenty of oil, and volatile salt whitening the whole Receiver, comes over; being that which no root or stalk can do; for the chiefest virtue, odor, and taste of vegetables, animals, and minerals is found in the seed, in which thing provident Nature has done very well, whilst she attributes the chiefest faculties to the seed, being more obnoxious to injuries then the rest, which is also preserved, nourished, and cherished by its like.

Now this discourse which might otherwise have been omitted, was therefore appointed, that the cause of the germination of vegetables might be made the more manifest; and that what things have been spoken of the attraction, and fixation of all things might the better be understood. The germination therefore, and multiplication of both minerals, vegetables and animals must be spontaneous, and not forced, as is that barren and frustraneous of the false Chymists, because preternatural. Wherefore when you fix anything be cautious in the adding of anything that should retain it, with which nothing can be fixed. Fire indeed does always do its office; but it knows not how to help any preternatural thing; which it does wholly destroy, against which nothing can be prevalent, unless it be rightly ordained according to Nature.

And thus, much is spoken for instruction sake, to you that intend to fix anything, lest otherwise you lose your labor.

Of the Making of the Best Crucibles,

The best crucibles that are requisite for the fourth furnace, not being found in every place, I thought it worthwhile to set down the manner of making them: for I am not ignorant how oftentimes many for want of these are constrained to be content with those that are useless, and truly with great loss of metals, whilst the crucibles are broken in the fire, and Consequently with a tediousness in drawing them out of the ashes.

Chymists have been in a great error a long time, and not only they but also goldsmiths, and they that separate metals, as also others that need the help of crucibles, who persuade themselves that the best earth that is fit to make the best crucibles is to be found nowhere but in HASSIA; and therefore with great charges have caused that Gibsensran crucibles be brought over; not considering that almost in every place in Germany such earth is to be found, which indeed is a very great folly of men, proceeding from the not knowing of good earth which is to be found almost everywhere. I do not deny but that the earth of HASSIA is very good for crucibles, tiles, retorts, and other vessels, which are to be set in a very great Fire, for which cause also is commended Gibsenian, and Waldburgensian crucibles.

    A few years since some have made their crucibles, and other vessels that will endure the fire well, of earth brought out of ENGLAND, and FRANCE into HOLLAND, which have retained metals very well in the fire, but not salts, because they are too porous and not so compact as those of HASSIA, wherefore those of HASSIA are still preferred before others, retaining better metals and salts. But although this earth be brought from thence to other places, yet such strong crucibles could not be made thereof, the cause whereof being not the constitution of the air, and place to which some have falsely imputed it, but an error in the making and burning of them. For in HASSIA there is a great abundance of wood, of which there is no sparing in the burning the crucibles even to the stony hardness, which could not be done by a small Fire of tursses.

    The like error is committed about stone pots, and other vessels which are made at FREEBEMIUM and SIBURGUS, and other places near COLEN, which are carried almost through all Europe, the goodness whereof is ascribed only to the earth, and not to the making. But now experience has taught us that any good earth does become stony in a violent fire, without respect of the place where it is taken. Wherefore it is very probable, being a thing possible, that such vessels are made elsewhere: for every earth being burnt retaining a white color, viz. with an indifferent Fire, makes pots, and crucibles porous, but with a stronger, and with a longer delay, compact like glass, especially if common salt be cast in a plentiful manner upon them, being burnt with a very strong fire, because it adds to them being very well burnt within an external glassy smoothness, by which means they will be the better able to retain spirits in the fire. Wherefore let no man doubt concerning the making the foresaid vessels of any other earth that is white in burning, with the help of a very strong fire: which by how much the greater whiteness it gets in burning, by so much the better and excellent pots it makes; and seeing there is a great difference of making crucibles to be set in the Fire, and of stone pots retaining liquid things, I shall show the manner of making both, viz. of stone pots belonging to the first and second furnace, and of crucibles to the fourth, and thus it is.

He that will try the goodness of white and pure earth, viz. whether it grows stony in the fire, let him cast a piece of crude earth of the bigness of a hens egg into a very strong Fire, observing whether it does quickly or slowly cleave and break in pieces; which if it does not cleave and become powder, although it may have some cracks, is good earth, and fit for burning, if the mixture be well made, in which lays the art.

The earth that is to be burnt, for pots, receivers, and bottles, needs no other preparation then that for bricks, which because for the most part it is too fat, you must mix with it clean sifted fusible sand, tread it with your feet, and knead it with your hands before vessels be made thereof; which being made are to be dried in the heat of the Sun, or in some other warm place; and being dried are to be burnt in a very strong Fire for the space of twenty four or thirty hours, on which in the mean time you may cast salt if you please, which being thus burnt do like glass retain easily all liquid things. But let him that makes crucibles, tiles, bricks, and other vessels appointed for a very strong Fire, use more diligence in the making of them. And truly first he must beat very small with a wooden hammer, the earth being dried well in the Sun, or elsewhere, and being beaten searse it through a great searse, and to one part of the sifted earth mix two, three, or four parts (the fatness of the earth being considered) of the earth burnt in a potters furnace, and powdered, which being mixed with a sufficient quantity of water he must tread with his feet, and afterwards knead with his hands, and the earth will be prepared for the making of vessels, and when he makes crucibles and tests, let him provide for wooden moulds both greater and smaller, made in a turn, by the help whereof they may be made, for the aforesaid vessels cannot be formed by the usual art of the potters; because the matter of them must be very lean, appointed for a most strong fire; wherefore commonly they are made by the help of moulds after the following manner.

    Let a piece of the prepared earth be applied with your hands to the mould, which you must hold in one hand, applying and fitting the earth thereto with the other, or hold it with your legs, that the earth may be applied with both your hands. Also you must first rub the mould very well with clean sifted sand, for else the earth will so stick to the wooden mould, that a crucible can scarce be taken off without danger, which being done, it is further fitted by striking it with a wooden instrument smoothed for the purpose, by which means the crucible lies very exactly upon the mould, for by this means crucibles are made very strong; which being done also let the crucible be taken off, and set upon a board, and be dried, first in the air, then by the heat of the fire, or sun, and then be burned in the first chamber of our fourth furnace, or in a potters furnace. And if you intend only to melt metals and not salts, you need not burn the crucibles if they be well, and exactly made.

Now this caution is to be observed in melting by the help of crucibles not burnt, that you must give fire above by little and little, for fear of breaking the crucibles feeling a sudden heat.

Now that they may be made equal in strength, weight, and thickness, you must weigh one crucible rightly made by the help of the mould in one scale, and a piece of the prepared earth, which is to be put into the other scale, and if they be equal in weight, take that piece out, and put in another; and this do so often, till you be come to the number of the crucibles which you would have made: By this means they are made equal, and you need not cut off any overplus of the earth when it is fitted to the mould, because all are made equal by reason of the equal weight of the matter of each of them, and the work is sooner done then otherwise,

This indeed is the best way but tedious and laborious, wherefore considering the matter a little more seriously, I found at last that the following way is far better than the former: whereby not only stronger crucibles are made, but also more in one hour, then in that former common way in three or four. Where first, the mould is made of latten (on which I advise you to apply the earth) signified by the letter A, viz. that being the best, which is made by the help of fusion. Then the counter—mould answering this, signified by the letter B, yet so that that do not enter too deep into this, not touching the bottom by the distance at least of one fingers breadth; but in greater crucibles a greater thickness of the bottom is required, as the practice will teach you.

Let him therefore that is making crucibles apply the earth to the mould, as has been above said in the first manner, which being done, let him again take off the crucible that is formed or cast, and set it in the air to be dried. Then having first made a sufficient number of crucibles, let him make the mould clean from the earth or sand, and anoint it with grease, or oil Olive taken up with a sponge, as also the counter—mould into which let him put the crucible being half made and dried, and into this mould, which he must strike above once or twice or thrice with a heavy wooden mallet, that the earth may be rightly and exactly applied to the mould; which being done let him take off the mould, and turn the countermould together with the crucible, which let him knock a little against the form (where the crucibles are made) and let him take in his hand the crucible falling from thence: which he must afterwards dry and burn, as has been above said in the first manner. And by this way are made the best, and the best proportioned crucibles, fixed and smooth, not only for melting of metals, but also for minerals and salts; the like to which I never yet saw, as being without all danger, if so be rightly made of the best earth. And that they may be made equal in weight and strength, they must be weighed as before has been said. And this labor is easy and pleasant, when they are made with one's own hand, and that greater or lesser at pleasure.

After the same manner, also are made tests, viz. by the help of the like kind of moulds, which must not be long but plain like shells as appears by the annexed Figure. A. and B. Not only tests but also cuples are made by the help of these moulds (See the fifth and sixth figures).

Now tests are made more easily this way then crucibles, because the earth only weighed, and being handled with the hands is put into the counter-mould, which you must with the upper-part press hard; that it may be made conformable to the mould, viz. plain, not long, that which may easily therefore be made; and for this cause those crucibles are easily again taken out, viz. if the mould be turned, or the counter-mould be a little knocked against the sides of the form. And if the earth be beaten in too fast that it goes out at the sides, you must cut it off with a knife, or else the crucible, or test is hardly taken out, sticking to the brims, which practice will teach you. For all things cannot be so accurately demonstrated by a pen.

And take this for a caution, that you do not make your tests and crucibles of earth that is too soft, but of that which is half dry, otherwise they are hardly taken out of the moulds; for that is more easily and rightly applied to the mould. And if you proceed rightly according to the prescript, scarce one crucible of a hundred will be lost.

This also is to be observed, that the superfluous earth which is cutoff must not be mixed again with the mass for crucibles, because it is spoiled with the fat, or oil that is smeared over the moulds, and therefore cannot be so well mixed again, and being burnt cleaves, for which cause bad crucibles are made. Wherefore it is to be kept apart for mending of furnaces that are spoiled with an extraordinary heat of the Fire; or for covers of crucibles that are to be made by the help of the hands only, or of moulds, which we cannot want, if we would work all things exactly.

Now for tiles, and other vessels that serve for distillation, and melting, they are made by the help of wooden moulds after this manner. Let the mould be made exactly like to the tiles, and other vessels, then cut off leaves from the earth being very well prepared, with a copper wire upon two equal tables of wood, and then a piece of the earth is to be laid with a knife upon the mould, that it may there get some hardness; which afterward is to be taken away, dried well, and burnt. And if anything further is to be done, viz. by cutting off, or adding, it must be done by earth half dried, or a little hardened. For by this means any one may get for himself earthen vessels that are necessary, without much cost or pains for certainty sake. For those that are sold, are negligently made, in which oftentimes in the drying, the cracks which are made, are filled up with some earthen liniment, before they are burnt, which therefore are not durable in the fire, but are broken, and that oftentimes with great loss of the metal, which is again to be gathered out of the ashes by the help of a tedious washing. It is better therefore to work those vessels with one's own hand for certainty sake. For not all and every crucible can always and everywhere be made equal, and be of a like durableness in the Fire, though they are made most diligently: and therefore, a consideration being had of their goodness, they may be used for divers uses, and the better may be used in the melting of the better metals. But let no man persuade himself that all these can indifferently hold in the Fire, although they be the best of all, how many so ever you make; for I never yet saw any earth which could hold litharge in the Fire and salt of Tartar, because the best that ever I saw is not free from penetration of them, which is the greatest impediment of some profitable operations, which therefore are omitted.

And let this which has been spoken, suffice concerning the making of crucibles: let everyone therefore that has a care of his business, use better diligence for the time to come in the making crucibles for more certainty sake, and he will not repent of his labor. Now how Tests and Cuples may be exactly applied to the aforesaid Molds, is not my work at this time to show, because many years since it has been done by others; especially, by that most ingenious Man, LAZERUS ERCKER, whose writings concerning the manner of making of Tests and Cuples I cannot mend, to which Authors I refer the Reader, where he shall find sufficient Instruction and Information concerning this matter. But there are also other Tests, of which I shall say nothing in this place, but elsewhere happily I may, by the help whereof, lead is bettered in trial if it be sometimes melted again.

Of the Vitrification of Earthen Vessels belonging to the first and second Furnace.

In the defect of glass Instruments belonging to our first Furnace, you may make such as are very useful, of the best Earth, which being well glazed, or double glazed, are sometimes better than old Glass; especially, those that are made of Earth that do not drink up the spirit, such as is found almost everywhere, which becomes stony being burnt:

    Now the Art of burning has not hitherto been so well known, of which something has been said already, where the Earth being burnt with a very strong fire, is made so compact, as that it becomes hard and solid as a stone. The Potters Furnaces being too weak for this strong burning, there is required, a peculiar Furnace for this Work; in which, the strongest fire for this Work; in which, the strongest fire for the burning of them may be made: But because nobody thinks to build such a one, only for some few Vessels not worth the spending of costs and labors: there is yet another way of vitrifying of any sort of Earth (red Clay only excepted) not to be slighted if well done; especially, if the matter vitrifying when it is cold after the burning is ended, does not cleave and chop, and it is not hurt by corrosive spirits as the glass made of lead, retaining spirits, as well subtle as corrosive, as that white vitrification of the ITALIANS and HOLLANDERS: you must therefore in defect of a fitting Furnace, wherein Vessels being burnt become stony, make them of the best Earth, and glaze them with the best Glass of Tin, but not of Lead; and by how much the more the calx of Tin goes into the vitrifying mixture, so much the better is it made; for Tin being reduced into a calx with Lead, has no more affinity with corrosive spirits; wherefore it is more fit for vitrification, But he that will not be at so much costs, let him vitrify with Venice Glass powdered, which vitrification also is not to be slighted, requiring a very great heat for the burning, and therefore flowing with great difficulty in these common Potters Furnaces; wherefore you must mix some Borax with the Glass, that it may flow so much the more easily in the Potters Furnace; else you must pour upon the earthen Vessels being burnt, Water mixed with Glass, so that the Glass may stick to them everywhere exactly, which afterwards being well dried, shall be gathered together into one heap artificially, lest they take up too great a space, like earthen Dishes that are to be burnt, and afterwards compass them round about everywhere with burnt Bricks, a hole being left open above for the casting in of coals, yet so, that the Bricks be distant from the Vessels the breadth of an hand, whereby the coals being cast above, may the more freely go round about down to the bottom; which space being filled with dry coals, you must put upon them other living coals, that the fire being kindled above, may by little and little burn downward and perform its work; which being done, the Vessels will be out of all danger, if so be they are all well dried.

The fire being kindled and burning, you must cover the hole with stones, until the fire of its own accord be extinguished; the coals being spent and the vessels become cold.

N. B. Now if there be a great heap of vessels, you must first, the coals being burnt, add fresh coals once more; for else the vessels being placed in the middle, cannot be sufficiently burnt, nor the glass sufficiently flow; wherefore caution is required in the governing of the fire in this manner, where, if all things are rightly done, the vessels are better and more truly burnt and vitrified than in any common Potters Furnace whatsoever; yet with greater danger to the vessels than in a Potters Furnace compassed about with walls. But let him that burns crucibles and other smaller vessels, burn them in our melting or distilling Furnace, being covered with coals, giving Fire first above, for so I myself was wont hitherto to burn all my crucibles, and burn and glaze all other distilling vessels; and this in defect of fitting Furnaces is the best way of burning and vitrifying, where in three or four hours' space, the vessels are exactly burnt and vitrified. Now the earth that is to be burnt quickly, must be the best, and durable in the Fire, for fear of breaking of some of the vessels. Let him therefore in this case for security sake, use our fourth Furnace, who has built it with his chambers, in the first whereof he may burn and vitrify without any danger. But that foresaid way of burning and vitrifying, is not to be slighted, wherefore I would have you be admonished to be cautious in giving of Fire, that you give no more or less than you should, lest afterwards you impute the cause of your error committed, to me, whilst the vessels are broke, as if I had not wrote the Truth, but to thyself that errs, and must for the future be more diligent, and cautious in this work.

    I know other vitrifications of divers colors hitherto unknown, and indeed most secret, not to be communicated to everyone indifferently: but he that knows how to reduce metals into a true glass, retaining the color of its metal, is indeed the inventor of a very great secret; to whom, if he considers the matter more profoundly, and exercise himself therein, a Gate is open, with the blessing of God, to a greater light.

    There are also other vitrifications, with which the earth being covered does appear, as if it were adorned with Gems; but because it is not our purpose now to treat of such kinds, I shall make an end of vitrifications, one only excepted, which I shall communicate for the sake of the Sick, and Physicians; and it is this:

Make little earthen Cups very smooth and white of the best earth being burnt: then make the following glass to flow in a very strong crucible, in which dip one cup after another, being held with tongs, and first made red hot in some little Furnace, letting them lye covered therein for a while, that the earth may the better attract the glass; which being done, let them be taken out, and be set again into the foresaid collateral Furnace, where they were before made red hot, when one is taken out, dip another in the molten glass in its place, which also is again to be set as the first into the foresaid Furnace; and this is to be reiterated so often, until all the pots be covered over with glass: all which being done, the Furnace is to be shut close everywhere, that the wind enter not into it, and so it is to be left until it become cold of itself, and the glass covering over the cups remain entire, which otherwise cannot be if the cups be set in a cold place; now the glass is made after this manner.

Take of crude Antimony two parts, of pure Nitre one part; grind them well, being mixt together, kindle the mixture being put into a crucible with a red hot iron, and the Sulphur of Antimony will be burnt together with the Nitre, a mass of a brown color being left behind, which you must take out while it is hot with a spatle that it may cool, which afterwards being melted in another strong crucible for the space of half an hour, or an hour, makes that glass with which the aforesaid cups with their covers are covered over.

Of the Use of the Aforesaid Cups.

There is no one that can deny that Antimony is the most excellent of all vomitives, wherefore, so many and so various preparations have been invented by Physicians for the taking away of the malignity thereof; whereof I have showed some, together with the use thereof in the First and Second Part of this Book, where always one is better than another; yet notwithstanding 'tis confessed, that Antimony reduced into Glass, is sufficient to purge the Stomach and Bowels from all corrupt Humours, and that without all danger, (being rightly administered) as well by vomit as by stool, by which means many grievous imminent Diseases are not only prevented, but also presently cured.

But you infer, that this is yet a crude and imperfect preparation, and therefore not so safe. To which I answer, the Antimony that purges, needs no preparation, for if all the crudity thereof were wholly taken away by fixation, it would no more cause vomiting or stools, wherefore the aforesaid glass of Antimony is not to be feared, because it is not dangerous, but may safely be given to Children that are one or two years old, but not in form of a powder, but in infusion or extraction of its chiefest virtue made with honey, sugar and wine, sweet or sour. After which manner being given, it attracts from all the bowels all vitious humours, and evacuates them as well upward as downward, without danger; of which thing elsewhere more at large. Let him that uses the aforesaid Cups, infuse one or two ounces of wine, and set them a whole night in some warm place, and the wine will attract from the glass so much as does suffice it, which afterwards being drank in a morning, does perform the same as an infusion made with the powder of Stibium; and this is a more delicate way than the other, because a Cup is sent to the Patient that he may infuse in it the space of a night, two or three spoonfuls of proper wine, placing it in some warm place, which he may drink up blood warm in the morning, with a due ordering of himself afterwards: Which, in my judgement is a more delicate way, being made with ones own wine, and ones own hand, than that tedious way of potions, both large, bitter, and nauseous. And this Cup may oftentimes be used, and if at length the wine should not attract sufficiently, the Cup with the wine is to be set in seething hot water for a little time, that the wine might the better attract, and work, when need shall require.

    Now he that gives such kind of Cups to others, must instruct them concerning the ordering, and administering the same. One Cup is sufficient for the Master of a Family, with his whole Family for all the days of their life. It is not to be used by all, and every one, and in all Diseases indifferently, but only by those that are strong and young; and where the principal parts are not hurt. Cups may also another way be covered over with Glass without Antimony, as follows.

    Sublime AURIPIGMENTUM, in a Glass or Earthen Gourd; and take the gallant golden colored Flowers thereof, which being after a peculiar manner melted, yield a red and most beautiful Glass almost like an Oriental Ruby, which being broken in places, may be used instead of an Ornament; but this is more soft, and brittle, than Glass of Antimony. This Glass, or those Flowers of AURIPIGMENTUM, which are not yet reduced into Glass, do notably glaze the aforesaid Cups with a red beautiful Color.

He therefore that will vitrify the foresaid Cups, must first heat them red hot in a Fire made with Coals; and being thus hot, dip them in the aforesaid melted Flowers, and being taken out thence, put them under an earthen, or Iron hot vessel, and there let them cool; which do perform the same things as those which are said of the Antimonial Cups.

These Cups are not dangerous, as to be feared, because as Antimony is corrected by calcination, so AURIPIGMENTUM is by sublimation: from which if all the malignity be taken away either by Fire, or by nitre, the vomitive virtue is taken away, as afterward shall be demonstrated more at large in these five parts, when they shall come forth again with enlargements, viz. what purging things are, and how they put forth their virtues, a consideration being had of their malignity.

There are also other ways of vitrification, and indeed very fine, and most desirable by all, if they should be communicated; but because it is not now my purpose to treat here of mechanical things, but only of some particular vitrifications of vessels belonging to our furnaces, I am resolved to omit them at this time, and make an end of these things. I am resolved, God willing, to set forth these parts more corrected, and in a larger manner, where many excellent things now omitted for some reasons, shall be published and communicated.

Wherefore I will now put an end to this fifth part, where although I might have added something that is singular concerning artificial furnaces, yet because time will not now permit, it shall be deferred to another time and place, where we shall treat further of the examining, trying and separation of metals: For the best way of melting of metals in a greater quantity has not yet been known: And although they that deal in minerals persuade themselves of the perfection of their art, yet I can demonstrate an easier, and more compendious way of melting of metals in a shorter time, in a greater quantity, and with less costs and pains. Of which more at large elsewhere, wherefore (Courteous Reader) be contented with these things, and if I shall see that these few things be acceptable to you, I will sometime hereafter for your sake and to your profit communicate WONDERFUL SECRETS which the world will not believe, and which hitherto are hid, either out of envy or ignorance.

FINIS.

An Appendix

Two years since I began to publish my new invented furnaces where also there was mention made of some secrets, which though I thought never to divulge; yet nevertheless I underwent many troubles for the communicating of them. Wherefore I beseech everybody that they would no more create troubles to me or to themselves by their petitions or writings, because for certain causes I shall for the future communicate nothing but those things which follow. Expect therefore patiently the time of another Edition, when these five parts shall come forth more corrected and enlarged, and many most choice secrets shall be communicated, which were for certain causes omitted in the first Edition.

I shall now God willing communicate those things which follow, yet upon this condition (because many are such, that by means thereof you may with a good conscience, without hurt to your neighbor, through God's blessing, get great riches) that you be mindful of the poor, and a good steward of riches got honestly, and use them to the glory of God and the eternal salvation of your soul.

The preparation of corn, as of Barley, Wheat, Oats, & etc. of Apples, Pears, Cherries, & etc. where fermentation being made they do yield by way of distillation a pure spirit very like to the spirit of wine without great costs; of the remainders whereof if the matter were corn, may be made good beer, or vinegar; but if the matter were any kind of fruit, as apples, pears, a very good drink like to wine, so that by this means you may find a double profit, by which you may not only have whereby to live honestly, but also to lay up for your heirs.

An excellent and wholesome drink of fruit, and corn, that is durable and like to Spanish, French, and Rhenish wine.

A distillation of the AQUA VITA of certain vulgar things not costly and like to the AQUA VITA of French and Rhenish wine.

A preparation of sugar like to the Western, and of tartar like to the natural Rhenish, out of honey and not costly; where one pound of sugar does not exceed the price of eight or ten stivers, and a pound of tartar, that does not exceed the price of two stivers.

A peculiar purification of crude tartar without loss, and a reduction of it into great crystals not costly, so as the price of one pound does not exceed six stivers.

The taking away of the ingrateful taste and odor of honey so as afterwards there may be made from thence a certain good AQUA VITA retaining no more the smell and taste of honey: also, a very good Mead or Methagline like unto very good wine, with which the same things may be done as with the best wine.

A preparation of Mead out of raisons, great and small, very like in all things to Spanish wine; out of which also is made a very good vinegar without great costs.

A preparation of wine and good vinegar of wild grapes.

Durable and wholesome drinks of gooseberries, barberries, strawberries, and the like.

The mending of troubled acid musty wines, & etc.

The preparation of a very good vinegar out of certain vegetables which are to be found everywhere, which may be compared to that which comes out of

France, and in a great abundance, whereof two rundlets of nine Gallons do not exceed the price of one Royal. (A Royal or Imperial is 4l. 6 d.)

The promoting of the ripening of wines of the cold countries of Europe (a very few that are very cold being exempted) that they may yield very good sweet and durable wines, whereas otherwise they could come to no maturity, being very like to those which hotter countries yield.

A certain secret way of carrying wines from mountainous places, where carts, ships, and other commodities are wanting, where the carrying of ten pipes, does not exceed the price of one pipe otherwise carried, so that by this means, outlandish wines may be brought to any place with great profit.

A very good and easy preparation of verdegrease out of copper, whereof one pound does not exceed the price of six stivers.

A new and compendious distillation of vinegar, of which a rundlet of eighteen gallons does not exceed the price of half a royal, with which many things may be done, especially the crystallizing of verdegrease, of which one pound prepared after this manner, does not also exceed the price of half a Royal.

A compendious and very easy way of distilling a very strong spirit of urine, and that without any cost and pains, so that twenty or thirty pints shall not exceed the price of one royal, being very excellent in medicine, Alchemy and Mechanique affairs, by the help whereof a most beautiful blue vitriol may be made out of copper, being very profitable in Alchemy and medicine, making silver so fusible, that by the help thereof, glass vessels, as basins, dishes, and candlesticks, & etc. may be so guilded as to be taken for silver.

A way of distilling the spirit of salt in a great quantity, and that with small costs, so that one pound thereof will scarce exceed the price of six stivers being very excellent in Alchemy, Medicine, and other Arts; especially for the doing of these following things, viz. the separation of gold from silver without hurt to the Cups or other things, also the solution and separation of gold mixt with copper and silver by the force of precipitation, where the MENSTRUUM that is preserved, may again be used for the same uses, which separation is the easiest of all other humid separations, whereby gold is reduced to the highest degree.

The separation of volatile sparkling gold out of sand, & etc. very profitable, without which otherwise it could never be separated, neither by the help of Washing, nor by Mercury, nor by the force of Melting.

An artificial secret, and hitherto unheard of, trying of stubborn Metals, finding out their Contents, which otherwise could not be found out: for oftentimes there are found golden mines, which are stubborn, in which nothing is found out by the common way, and therefore they are left unlabored in, and sometimes elsewhere, where there are not found Mines of Metals, there are found other things, as white and red talc, that yield nothing, being tried the common way, or very little, all which yet abound with gold and silver, which may be separated this way.

A new and unheard of compendious way of melting Mines in great plenty, where, in the space of one day, by the heat of a Certain Separating Furnace, more may be melted than by the common way in the space of eight days, where not only costs are saved, but also is hope of greater gain.

Another way for the better proving of things melted, and a new way of separating silver from lead.

A very speedy way of melting Minerals, whereby they are melted in great plenty, by the help of Pit-coals in defect of other coals.

The fixation of Minerals, Sulphureous, Arsenical, Antimonial; and others that are volatile, which cannot be retained and melted by the force of fire, by the help of a certain peculiar furnace with a grate, so that afterwards they may by infusion yield gold and silver.

The getting of gold and silver, that sparkles, and is rarified, out of sand, pure clay, flints, & etc. by the help of melting.

The separation of gold lying hid in baser minerals and metals most profitable, which cannot be done the common way.

A very quick Artificial and easy separation of melted gold and silver by the help of fusion, so that in the space of one day, by the help of one furnace, some hundreds of Marks may be separated with far less costs and labor, than by the common way by cement and AQUA FORTIS.

The reduction of elaborated gold, of chains and other ornaments into the highest degree; also, the separation of gold from guilded silver, by the help of fusion, by which means a hundred marks are more easily separated than twenty of the common way.

A certain way whereby more silver is separated from lead then by the Copper.

A separation of good gold from any old iron, which although it be not a labor of great gain, yet it is sufficient for those who are contented with a few things.

A separation of gold and silver, from tin or copper, according to more or less. The maturation of mines, so that they may afterwards be able to yield more gold and silver, then by the common way, also the separation of gold and silver out of Antimony, Arsenick, and AURIPIGMENTUM.

The separation of the external sulphur of VENUS, that the Son CUPID may be born.

The separation of silver from the cuples, into which it enters in the trial without melting or any other labor or cost.

The preparation of divers earthen things to be done in any part of the world, like to the Porcelain, that hold fire and retain spirits.

A certain Allome exalting and fixing any color, especially requisite for scarlet and other precious colors, with a certain perpetual cauldron, that does not alter colors, and is not costly.

A making of colors for painters, as of purple gum, ultra-marine, not costly, and especially of that rich white, never before seen, like to Pearl and Margarites; also, a peculiar coloring of gold and silver.

To conclude, I refer the Reader unto the Residue of my Books, that treat of those Secrets more plainly; which I am resolved shortly to put forth.

Those Secrets are all openly taught in the following Treatises, as in the Explication of **Miraculum Mundi**, **Apology against Farner**, **Prosperity of Germany**, & etc.

FINIS.

A Word from the Publisher

Thank you for purchasing this small work from The R.A.M.S. Library of Alchemy. During his lifetime, Hans Nintzel was dedicated to the identification, acquisition, study, retyping and, when necessary, translation of what he considered to be the most important known works on Alchemy. Hans was assisted by his sparse network of fellow Alchemists, all members of the Restorers of Alchemical Manuscripts Society (R.A.M.S.). I was an active member of R.A.M.S.

My goal is to publish all of the works originally made available through R.A.M.S. as photocopies. To facilitate this, I have chosen to have the books professionally printed. I also have a few titles that I intend to add to the original R.A.M.S. Library, selected by strict criteria established by Hans.

The works from the original R.A.M.S. Library are republished by R.A.M.S. Publishing Company in the collection, "The R.A.M.S. Library of Alchemy," with permission of the Estate of Hans W. Nintzel.

If you have a work on Alchemy that you believe should be a part of the R.A.M.S. Library, please contact me through R.A.M.S. Publishing Company.

Philip N. Wheeler

The R.A.M.S. Library of Alchemy

The study and practice of Alchemy was extremely important to Hans W. Nintzel. He assembled this Library over a period spanning more than three decades, guided by his teacher Frater Albertus. The R.A.M.S. Library of Alchemy includes all of the most valuable Alchemical texts that Hans painstakingly located, acquired, retyped, and translated during his lifetime, with help from other R.A.M.S. members.

The following is a list of the volumes that are currently available. Volumes that contain works from multiple authors may have only the principle author or editor listed.

Volume	Title	Author or Editor
1	Twelve Keys of Basilius Valentinus	Basilius Valentinus
2	Triumphal Chariot of Antimony	Basilius Valentinus
3	His Secret Book	Artephius
4	The Golden Work	Hermes Trismegistus
5	Three Works of Ripley	George Ripley
6	Four Works of Paracelsus	Paracelsus
7	Bacstrom's Notebooks, Part 1	Sigismund Bacstrom
8	Bacstrom's Notebooks, Part 2	Sigismund Bacstrom
9	Summa Perfectionis	Geber (Abu Musa Jabir ibn Hayyan)
10	The Five Centuries	Rudolph Glauber
11	The Greater and Lesser Edifyer	Johann Grashoff
12	Chemical Secrets and Experiments	Sir Kenelm Digby
13	The Turba Philosophorum	Arisleus
14	Das Aceton	Christian Becker
15	The Art of Distillation	John French
16	Non-Violent Destruction of the Atom	Nintzel & Wheeler
17	Philosophical Furnaces	Rudolph Glauber
18	TBD	
19	TBD	
20	TBD	
21	Alchemical Symbols, Third Edition	Hans W. Nintzel and Philip N. Wheeler

22	The Book of Formulas	John Hazelrigg
23	18 Short Tracts	Hans W. Nintzel
24	Bacstrom's Notebooks, Part 3	Sigismund Bacstrom
25	A Discourse on Fire and Salt	Blaise Vignere
26	The Mineral Work	Johan Hollandus
27	The Vegetable Work	Johan Hollandus
28	Lamspring's Process	Lamspring
29	The Book of Abraham the Jew	Abraham Eleazar
30	Five Short Works of Glauber	Johann Glauber
31	The Metamorphosis of the Planets	Johannes Monte-Snyder
32	Four Works of Roger Bacon	Roger Bacon
33	The Golden Chain of Homer	Homerus, Kirchweger, Nintzel, Wheeler
34	Alchemy Rediscovered and Restored	Archibald Cochren
35	Aurifontina Chymica	John Houpreght
36	The Golden Fleece	Salomon Trismosin
37	The Transmutation of Base Metals into Gold and Silver	David Beuther
38	Sanguis Naturae	Christopher Grummet
39	A Revelation of thye Secret Spirit	Giovanni Lambi
40	The Holy Guide, Part 1	John Heydon
41	The Holy Guide, Part 2	John Heydon
42	Secreta Alchymiae	Kalid Persica
43	The Golden Treatise of Hermes	Hermes Trismegistus
44	Potpourri of Alchemy, Part 1	Hans W. Nintzel
44	Potpourri of Alchemy, Part 2	Hans W. Nintzel
46	TBD	
47	Selected Chemical Universal and Particular Processes	Alexius von Ruesenstein

http://ramsalchemy.jimdo.com

www.ingramcontent.com/pod-product-compliance
Lightning Source LLC
Chambersburg PA
CBHW081139180526
45170CB00006B/1850